D0572193

Workbench
Guide

to

Electronic
Troubleshooting

Workbench Guide

to

Electronic

Troubleshooting

Robert C. Genn, Jr.

Parker Publishing Company, Inc., West Nyack,N.Y.

© 1977, *by*

PARKER PUBLISHING COMPANY, INC.

WEST NYACK, N.Y.

All rights reserved. No part of this book
may be reproduced in any form or by any
means, without permission in writing from
the publisher.

Library of Congress Cataloging in Publication Data

Genn, Robert,
   Workbench guide to electronic troubleshooting.

   Includes index.
   1. Electronic apparatus and appliances--Maintenance
and repairs. I. Title.
TK7870.G422   621.381'028   77-1143
ISBN 0-13-965228-0

Printed in the United States of America

Illustrated by

E. L. Genn

How This Book Will Help

the Electronics Technician

This *Workbench Guide to Electronic Troubleshooting* will be a constant, reliable companion you will use frequently on the job. Conveniently bound in loose leaf form that allows for easy referral and handling while working, it is written for the technician who is searching for fast, practical and more economic ways to troubleshoot and maintain electronics equipment. Because of this, the only basic theory included will deal with fundamentals underlying some of the troubleshooting techniques that are not generally known.

Mathematics has also been kept simple so that all calculations can be done on a pocket calculator. Furthermore, emphasis has been placed on *work-related problems,* and many of the examples are from on-the-job experience by the author and co-workers.

Every chapter stresses inexpensive, simplified techniques that can be used for years to come. For instance, how to use resistors in place of expensive transformers is explained in Chapter 1. As another example, Chapter 2 shows how to make simple in-shop frequency checks having a tolerance of 0.02 Hz using common 60 Hz house current. To make fast repairs as easy as possible, all chapters will tell you *exactly* what to do in case of a trouble. As an illustration, Chapter 3 tells you that if your transistor oscillator will not start, try changing the bias voltage approximately plus or minus 20 percent of the recommended value.

Many procedures and simplified techniques deal with troubleshooting solid state components, and each step is explained in detail. Because vacuum tubes continue to have important applications in the electronics communications field, you will find many useful troubleshooting techniques for high power and cathode ray tubes used in video presentations. Frequently, speed is of the essence when repairing this equipment; and consequently, a step-by-step procedure is given in each case.

No workbench would be complete without the information contained in Chapter 13. A collection of essential information frequently needed is organized and coordinated in this chapter for quick, easy reference.

Never before has technical knowledge been so well rewarded as it is today. Higher pay, more prestige, and greater security—they all go to those who are successful on the job, and every chapter in this book provides you with the practical techniques that are essential to your increasing success as a technician.

Robert C. Genn, Jr.

Contents

Workbench
Guide

to

Electronic
Troubleshooting

Practical Troubleshooting Techniques for Calibrating and Repairing Electronic Instruments

The way a technician works with his instruments is probably as unique as his fingerprints. The distinguishing factors are called the *tricks of the trade*. This chapter includes tricks that are essential for correct meter readings and making new calibrated meter scales, as well as a quick way to calibrate analog instruments using an easy-to-find standard with a tolerance of $\pm$ 0.002.

Signal generators with a metered output are expensive. Right? Not necessarily. Why not build your own calibrated attenuator? It's easy. A few resistors and a VTVM are all you will need. Furthermore, as startling as it may sound, this same inexpensive resistor circuit can be used to meter voltage, current, and power outputs. These are just a few of the troubleshooting procedures and techniques that will be demonstrated in this chapter.

How to Eliminate Errors When Measuring Different Waveforms During Troubleshooting

Remember when the standard signal generator's outputs were a sine and square wave? Most of us knew what to do when using these waveforms. However, as usual, they goofed things up and brought out the function generator which included a triangular wave. (Translation:

"goofed things up" means that function generators were better and more sophisticated and I had to learn something new.)

Here's the problem: As long as the output of your function generator is a sine wave, your rms reading voltmeter is telling the truth. But place those probes across the output with the function switch in either the triangular or square wave position, and—don't you believe it.

To illustrate, if your meter is constructed around a peak reading meter scaled to read rms (this is done in the factory) and you place the test probes on the output of a function generator set at a peak of 5 volts sine, square, or triangular, it will read 3.535 volts. The truth is that only the sine wave output is a correct reading. The true reading for the square wave is 4.99 volts (rms) and the triangular wave is 2.88 volts rms.

Before we go any further, let's emphasize that very few shop instruments are actually rms meters. More than likely, the meter is a peak or average responding meter with the scale calibrated to read rms. What this means is that you must use correction factors to determine the true readings and, furthermore, they are different for each type instrument.

How Peak Reading Meters Scaled to Read rms Go Wrong and What to Do About it

Let's begin by listing the correction factors you need to use on a peak reading meter that is scaled to read rms. Table 1-1 clearly points out the hazards you will encounter unless you use the correction factor shown.

Take the square wave shown in Table 1-1 for an example. Notice that the correction factor is 1.414 times the actual meter reading. In this case, the meter reading shows that the square wave output of the function generator is 7.07 volts. Now, multiplying 1.414 times 7.07 produces 9.996 volts, or approximately 10 volts. In other words, the true square wave output is 10 volts rms.

Correction Factors You Will Need for Average Responding Meters Scaled to Read rms

Another common system is for the manufacturer to use an average

WAVEFORM	VOLTAGE VALUE			
	PEAK VOLTAGE	ACTUAL METER READING (AR)	AR x CF = TRUE READING	CORRECTION FACTOR (CF)
[sine waveform, 10 / 0 / −10]	10	7.07	7.07	1
[square waveform, 10 / 0 / −10]	10	7.07	10	1.414
[triangle waveform, 10 / 0 / −10]	10	7.07	5.76	0.815

Table 1-1: Correction factors for a peak reading meter scaled to read rms

21

WAVEFORM	VOLTAGE VALUE				
	PEAK VOLTAGE	AVERAGE VOLTAGE	ACTUAL METER READING	AR x CF = TRUE READING	CORRECTION FACTOR
	10	6.38	7.07	7.07	1
	10	10	11.1	10	0.9
	10	5	5.55	5.77	1.04

Table 1-2: Correction factors for an average reading meter scaled to read rms

22

responding meter and scale it to read rms, using a sine wave for calibration. The correction factor for this type of meter is shown in Table 1-2.

In reference to the square wave, a 10 volt peak square wave has an average value of 10 volts. But the reading on the meter will show 11.1 volts. This is because all the values were multiplied by 1.11 when it was calibrated at the factory. This means that you must mentally multiply by a correction factor of 0.9 in order to arrive at the true rms voltage. In the same way, the scale reading for a triangular wave must be multiplied by a factor of 1.04 to know the actual reading.

One other thing you should watch—Harmonics will affect the operation of these meters. In this respect, the peak responding meter is worse than the average responding type. Because of this, it is advisable to monitor the voltage under test for harmonics if accuracy is important. A simple, easy-to-make harmonic analyzer is described in Chapter 2.

Guidelines for Calibrating Meter Scales

Sometimes during our work, we find that we have to extend the range of a meter by using a voltage multiplier or current shunts. And, after the range is extended, many technicians like to construct new scales and glue them on the meter face. There are two common types of meter scales that we all have to use. These are the linear and the square law type. Figure 1-1 points out the difference in calibration points.

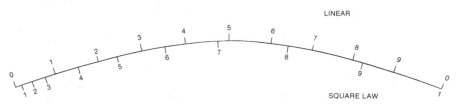

Figure 1-1: The linear and square law meter scales

How to Add a New Linear Scale to Your Meter

Suppose you want to make a scale for a linear meter. *Note: This is the familiar D'Arsonval meter.* To begin with, the deflection in this type meter is proportional to the current. Therefore, we can set up an equation relating to deflection and current, which is

$$\theta \propto I \qquad \text{or} \qquad \theta = K I$$

I	0.1A	0.2A	0.3A	0.4A	0.5A	0.6A	0.7A	0.8A	0.9A	1.0A
θ	10°	20°	30°	40°	50°	60°	70°	80°	90°	100°

Table 1-3: Calibration points for a linear scale

I	0.1A	0.2A	0.3A	0.4A	0.5A	0.6A	0.7A	0.8A	0.9A	1.0A
θ	1°	4°	9°	16°	25°	36°	49°	64°	81°	100°

Table 1-4: Calibration points for a square law scale

where θ is the angular deflection in degrees, $\propto$ means proportional to, and **K** is a proportionality constant.

To keep things simple, let's set up a maximum deflection of 100 degrees for a current of 1 amp. Now, using the formula $\theta = $ **K I**, we have $100 = $ **K I** and **K** works out to be 100. With this information, everything becomes easy. For example, $\theta = (100)\ (0.1\ A) = 10$ degrees. The calibration points on your scale will be as shown in Table 1-3.

How to Add a New Square Law Scale to a Meter

To calibrate a square law scale is just as easy. Since the deflection is proportional to the current squared, you can use the following equation with the definitions being the same as before.

$$\theta \propto \mathbf{I}^2 \qquad \text{or} \qquad \theta = \mathbf{K}\,\mathbf{I}^2$$

Again, let's assume a maximum deflection of 100 degrees at 1 amp. Doing this, we have a value for **K** of 100. This means that for a current value of 0.4, this angle is $\theta = 100 \times 0.4^2 = 16°$. Table 1-4 shows all the calibration points for a square law scale.

A Quick Solution to the Standards Problem

How many times have you tried to measure a standard 1.5 volt dry cell with a VOM and had to use the 10 volt range? You quickly found out that you couldn't read the scale with any degree of accuracy. The solution to this problem is to use a digital multimeter (dmm). With a dmm, you will use the 1 volt range. The dmm will sense that the voltage is too large and will light up the left-hand digit 1 (the so called half-digit). The reading will be 1.470 if you are using a 3 1/2 digit instrument, which is an impossibility with an analog volt ohmmeter.

It is well known that the accuracy of a common VOM is usually specified as a percentage of the full scale reading, generally 2 to 5%. However, the accuracy of a dmm is specified as $\pm$ a percentage of the reading plus one digit.

What does this mean? It means that most of your analog instrument calibration problems are over. For example, a reading of 1 volt would have a possible error of $(1.00 \times 0.001) + 1$, or a beautiful tolerance of $\pm$ 0.002 volts. The next time you are working in the shop and want to calibrate an analog meter, just pick up the dmm and your work will be over before you know it.

Troubleshooting Digital Instruments

The multimeter is the work horse around any shop. In fact, it is the principal—and sometimes only—test instrument most technicians use. So, if the digital multimeter develops a trouble, it could mean all jobs come to a halt. To make matters worse, many technicians back off from tackling anything dealing with digital circuits. The reason is that digital circuits have timing sequences and hard-to-catch pulses. The problem is that these pulses occur one shot at a time and the pulse width will vary anywhere between 0.5 μ seconds up to about 200 milliseconds.

How do you work on these circuits? Well, there are a couple of ways. One partial solution is to be rich enough to have a high speed trigger sweep oscilloscope in the shop. The other way (much less expensive) is to arm yourself with one of the several types of digital probes now on the market.

Troubleshooting with a Digital Probe

Perhaps the best low cost digital probe is the four-channel, hand held type with full memory capabilities. However, there are several other types that use three light emitting diodes (LED), are much less expensive, and are very good but not quite so versatile. Most of these probes use LEDs to indicate a logic state with a lit LED meaning logic 1 and a dark LED for logic 0. Figure 1-2 illustrates one type of a hand held, four-channel digital probe that can be used to troubleshoot most digital problems.

Where to start? Well, first take a look at Figure 1-3, a typical block diagram of a digital multimeter.

Notice that troubleshooting the upper section is very similar to what you have always done with the standard shop multimeter. But, much of the lower section of the block diagram requires an oscilloscope (two or three channel is best) or a logic memory probe. Of course, it is also

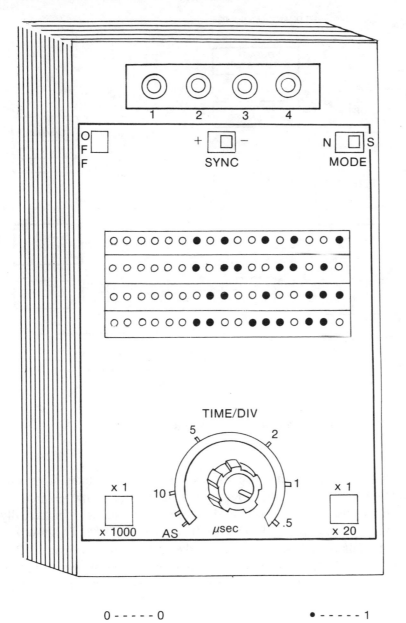

Figure 1-2: Four-channel digital logic memory probe

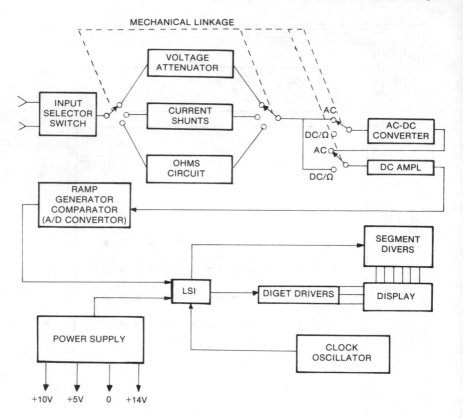

Figure 1-3: Digital multimeter block diagram

necessary to have a circuit schematic showing the logic states or get this information from the manufacturer for the instrument being repaired.

By using a schematic that shows the logic states, it is actually easier to use a digital probe than an oscilloscope or VTVM. To show you how simple this is, let's say that you use the probe shown in Figure 1-2 to measure four different points in the digital multimeter shown in Figure 1-3. Now remember, there are only two states—on and off. Your display would be something similar to what is shown in Figure 1-2 on this particular probe.

If this shows to be correct on the schematic, move on to the next test points. When you find an improper logic state, this is the defective stage. Sound easy? It is. Considering all the wave forms you must observe to troubleshoot a TV set, you'll find it's more like child's play.

How to Analyze Logic Circuits with a Scope

If you attempt to troubleshoot a logic circuit with an oscilloscope, it's important that you realize many common logic circuits operate at speeds that are high compared to the bandwidth of some oscilloscopes. For example, in a TTL circuit, a 15 to 20 nanosecond pulse is enough to trigger a flip-flop and you couldn't even see it on an oscilloscope with a bandwidth of 10 MHz. In fact, to see it properly you should have a scope with 50 MHz or better bandwidth.

When attempting to observe pulses in logic circuits, an X 10 (low capacitance) probe should be used. Place the probe ground clip as close to the point of measurement as practical and keep your ground lead short. A word of caution: It's possible to disable circuits temporarily (for example, flip-flops) by making direct measurements with coaxial cable. This is due to the excessive capacitance presented by the cable. Therefore, it isn't wise to use a scope to check out logic circuits without a probe. Furthermore, the compensated X 10 probe can introduce considerable error when you are working with high frequencies if you don't adjust it prior to use. Even if you move a probe from one channel to the other on a dual trace scope, you should be sure and readjust the probe.

Finally, unfortunately there are many times when even the best oscilloscope will not serve as a troubleshooting aid when working with very short time span pulses and the repetition rate is low. A possible solution is to use a storage oscilloscope. Or, it may be possible that you can modify the circuit you're working on so that you can temporarily speed up the repetition rate fast enough to make it observable on the scope.

How to Construct a Calibrated Attenuator
Using Nothing But Resistors

Many times a technician needs a metered voltage during repair of instruments or other electronic gear. Some signal generators have calibrated outputs but there are a lot that don't. If yours is one of these, all you need are a few resistors and a VTVM and you can get metered output right down to the microvolt range.

The problem boils down to providing a proper impedance match between the generator and equipment under test and, at the same time, attenuate the output of the generator to the desired level. One of the simplest ways to do this is to use an unbalanced T-type pad. A T-type pad schematic is shown in Figure 1-4.

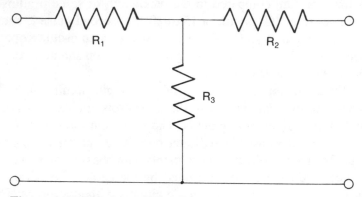

Figure 1-4: Unbalanced T-pad

To calculate the values of resistors R_1, R_2, and R_3, you can use the following formulas.

$$R_3 = 2 \sqrt{(N)(Z_{in})(Z_{out})} / N - 1$$

$$R_1 = Z_{out}(N + 1) / (N - 1) - R_3$$

$$R_2 = Z_{in}(N + 1) / (N - 1) - R_3$$

Where N is equal to the ratio of the power you want to lose in the attenuator to the power you want to deliver to the load. It's better to calculate the value of R_3 first because you will need this value to calculate the values of the other two resistors. If the entire test setup has matched impedances, R_1 and R_2 are equal in value.

Here's how it's done: Suppose that you want to attenuate a certain signal by a power ratio of 1/25, (14dB), and all impedances are matched at 600 ohms. The values of the three resistors you will need are

$$R_3 = 2 \sqrt{\frac{(0.04)\,(600)\,(600)}{0.04 - 1}} = 245 \text{ ohms}$$

$$R_{1,2} = 600\,(0.04 + 1)\,/\,(0.04 - 1) - 245 = 405 \text{ ohms}$$

Connecting the resistors, you will have an attenuator network like the one shown in Figure 1-5.

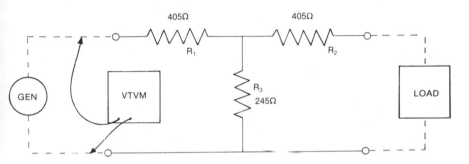

Figure 1-5: 14dB attenuator designed to work in a 600 ohm system

Now that you know the value of the resistors for the pad, it is easy to produce a metered voltage output. Let's say that you set the generator output at 1 volt. This would mean that the current has to be 1/600 A. Why? Because Ohm's law states that with 1 volt being applied to 600 ohms, it will produce a current of 1/600 A. The voltage across the parallel circuit composed of the 245 ohms in parallel with 1,005 ohms is 1 volt minus the voltage drop across the 405 ohm series resistor. This is

$$1 - (1\,/\,600 \times 405) = 325 \text{ millivolts}$$

In other words, you would have a voltage of 325 millivolts being applied to the equipment under test. Furthermore, make up three or four of these and you have a step attenuator. See Figure 1-6.

One last point—If you can afford to lose a small amount of signal, a T-pad will serve as an impedance matching device and save you the expense of a transformer.

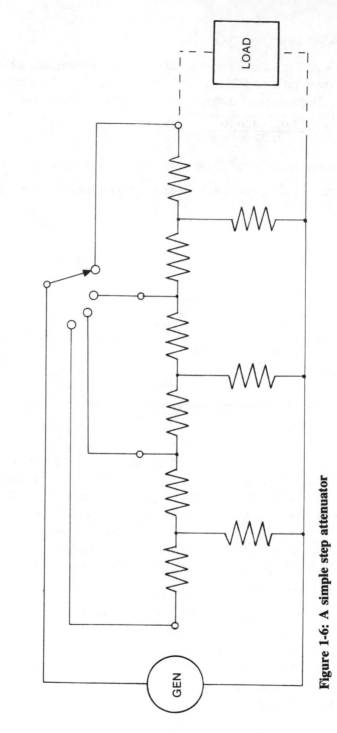

Figure 1-6: A simple step attenuator

Selected Electronic Measurement
Techniques Needed for Troubleshooting

How many times have we all needed to know the internal resistance of a meter? You will find a simple solution to the problem in this chapter. But that's not all. You will also find practical methods of using the commercial power at your workbench to measure frequency, capacitance and inductance.

Another practical technique we all need from time to time is how to make a quick dynamic impedance measurement. How to master all these measurements (and more) is explained as you read on.

How to Measure the Internal Resistance of a Meter

To extend the current range of an ammeter, all you have to do is shunt the meter with the proper value of resistance. The catch is . . . to determine the proper value of the shunt resistor, you must know the internal resistance of the meter. The following is a fast, painless way to measure this value. Figure 2-1 shows the circuit connections.

Start out with only the meter, potentiometer R_1, and power source. *Note: Any stable DC supply will work as long as you keep it near 1.5 volts.* Before you connect the DC supply, be sure and set R_1 to a high enough resistance value to limit the current to a safe value. To be on the safe side, use at least 100 megohms. For example, using a one and one-half volt supply, this produces a current of 15 μ A.

Now, carefully adjust R_1 until you have a full scale reading. Next, place a potentiometer of several thousand k ohms (set at maximum resistance) as shown in Figure 2-1. This is the potentiometer R_2. Adjust R_2 until the meter reads exactly one-half scale.

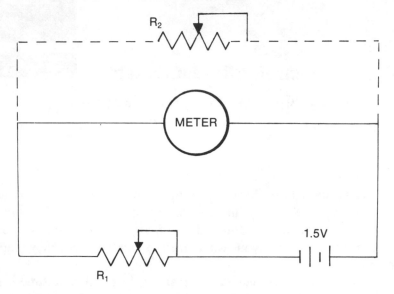

Figure 2-1: Measuring meter resistance

The next step is easy. Remove R_2 from the circuit and measure the resistance. By the way, if you happen to have a resistor decade box, the last step is somewhat easier. Substitute the box for resistor R_2, make the necessary adjustments, then simply read the correct value off the dial settings.

It may occur to you to ask, "Why not use an ohmmeter to measure meter resistance?" Unless you know how much current your ohmmeter will produce on each scale with the probe tips shorted and the voltage between the probes on each scale, it is strongly suggested that you don't try it. Incidentally, if you do not know these values, you should find out because it will save you a lot of burned-out semiconductor components in the future. Chapter 3 tells you why this is true and shows a successful technique for making ohmmeter checks on sensitive meters and semiconductors.

Using Your Oscilloscope to Check
Unknown Frequencies

Most of us are familiar with using radio stations WWV and WWVH to calibrate equipment using their broadcasts of 2.5, 5, 10, 15, 20, and 25 MHz. However, the following methods will extend the use of these frequencies by about 20 additional points. Furthermore, it is possible to extend the use of any other available standard by approximately 20 new points. Take the case of the 60 Hz commercial power being supplied to your workbench. It can provide calibration points from 6 to 600 Hz with a tolerance ± 0.02 Hz. How? Easy. All you will need is an oscilloscope and a 6.3 step down transformer plus the audio oscillator you want to calibrate.

First, place the scope sweep selector into the external sweep position. Next, plug the primary of the transformer into a 115 VAC bench receptacle and connect the 6.3 VAC output to the horizontal input of the oscilloscope. Now, connect the output of the audio oscillator to the vertical input.

HORIZONTAL

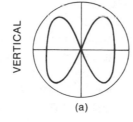

(a)

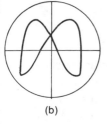

(b)

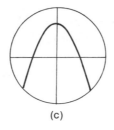

(c)

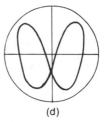

(d)

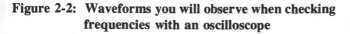

Figure 2-2: Waveforms you will observe when checking frequencies with an oscilloscope

Start off by tuning the audio oscillator until you see one sine wave on the scope. This indicates a frequency ratio of one-to-one, as we all know. However, tune the audio oscillator up to exactly 120 Hz and you will see the pattern shown in Figure 2-2 (a).

As you approach the 120 Hz point, you will see each of the patterns shown in Figure 2-2. But, the pattern you are interested in is (a) because it tells you that the unknown frequency is exactly two times as high as the known frequency. When you see three points along the top of the screen, the frequency is 180 Hz; four points represents 240 Hz, and so on. The way to determine the ratio of the two frequencies can be stated as follows:

$$\frac{\text{frequencies applied to the horizontal input}}{\text{frequency applied to the vertical input}} = \frac{\text{number of points counted on vertical}}{\text{number of points counted on the horizontal}}$$

The next logical question is, "What if I place the unknown frequency into the vertical input and the known frequency into the horizontal input?" The answer to this question is that you will see some submultiple of the unknown: 1/2, 1/3, 1/4, etc.

All of this means that you can use any of the many standard frequencies that are available around the world and calibrate either above or below the standard by about ten steps. However, ten-to-one is about the best you can do because it becomes difficult to interpret the oscilloscope presentation. For instance, using WWV's transmission of 400 or 600 Hz will give you up to about 4,000 and 6,000 Hz. Above these points, the screen becomes very muddy.

How to Construct and Use an Accurate Resistance Device

In this modern world, it is becoming more important each day to measure resistance in very small values to a high degree of accuracy. Although common ohmmeters have the advantage of speed and compactness, they lack the high accuracy of the null meter. The only limitation to accuracy and value of resistance you can measure with the following null method is determined by the components you choose to use. Another

beautiful thing is that it does not take all day to construct it. A schematic
diagram is shown in Figure 2-3.

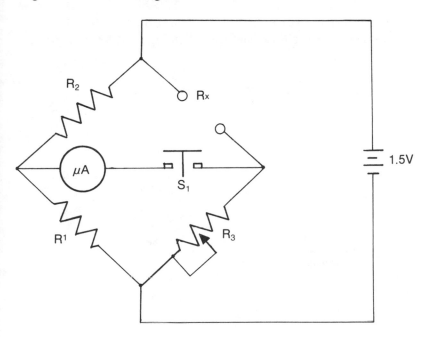

**Figure 2-3: Balanced bridge used to measure an un-
known resistance**

Let's assume you want to measure the resistance of a meter shunt.
To start, you would want at least one percent tolerance resistors for R_1
and R_2 and both of equal value. In a case like this, it is easier to use a
resistor decade box for R_3 because they are convenient and have the
desired accuracy. Also, the microammeter should be a zero at center
type scale.

In making the measurement, place the unknown resistance across
the terminals shown at R_x and adjust the resistor decade box for a zero
reading on the microammeter. Press the switch S_1 on and off rapidly
during the adjustment until you have an on-scale reading. Once you are
on scale, slowly bring it to zero. All that is left to do is read the resistance
off the resistor decade box.

Simple Techniques for Measuring Capacitance
and Inductance

You can use a resistor decade box and a filament transformer to check the value of a capacitor or inductor. See Figure 2-4 for the setup.

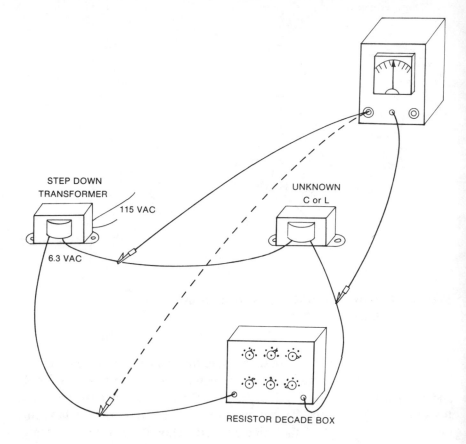

**Figure 2-4: Place your capacitor or inductor at the
point marked C or L.**

To find the value of an unknown capacitor, place the resistor decade box in series with the capacitor, as shown in Figure 2-4, and excite the circuit with the 6.3 AC voltage. Now, adjust the resistance until you measure the same voltage across both the resistor and capacitor. Read the value of the resistance; this is equal to the reactance of the capacitor.

The final step is to calculate the value of the capacitor by using the formula $C = 2.65 \times 10^{-3} / X_C$. For example, if the resistor decade box reads 5 k ohms, $C = 2.65 \times 10^{-3} / 5,000 = 0.53 \mu$ fds.

With the same setup that was just explained, here's how to determine the value of an inductor. However, in this case you want to use the formula $L = X_L / 376.8$. Let's say that you have a reading on the resistor decade box of 8,000 ohms. Simply divide 8,000 by 376.8. Your calculations should give you 21.23 . . . henrys. Watch these formulas because they are only good provided that you use 60 Hz line current to excite the test circuit.

How to Make and Use a Harmonic Tester

If you merely want to determine the total harmonic output as distinct from each individual frequency of the output, this isn't too difficult. Just suppress the fundamental frequency and measure the part that remains. A simple method to suppress the fundamental frequency is shown in Figure 2-5.

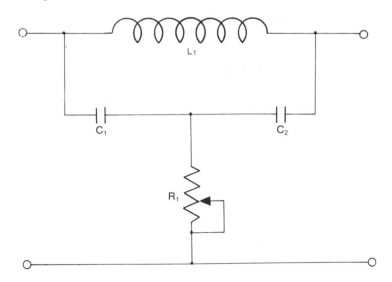

Figure 2-5: A bridge T-circuit

When you choose the components, a good rule of thumb is to start by making the reactance of the inductor at least ten times as high as the capacitive reactance at the fundamental frequency. After you decide on the inductive reactance value, use the following formulas to calculate the inductance or capacitance needed. These calculations are easily done on any four-function calculator.

$$C = 0.159 / f\, X_C \qquad\qquad \text{and} \qquad\qquad L = X_L / 6.28\, f$$

We'll assume that you want to check the harmonic output of an audio amplifier using a thousand cycle test tone. To make it easy, let's set the inductive reactance at ten thousand ohms. Now, calculate the value of inductance as follows:

$$L = 10,000 / (6.28)\, (1,000) = 1.59 \text{ henrys}$$

With these values, we can set the capacitive reactance at 10 ohms, which results in $C = 0.159 / (1,000)\, (10) = 15.9$ microfarads. Since it is always better to use a large capacitance value for a bypass capacitor, a 20 microfarad capacitor would work well here. Incidentally, the potentiometer R_1 is merely an adjustment for passing the desired amount of the fundamental frequency to ground.

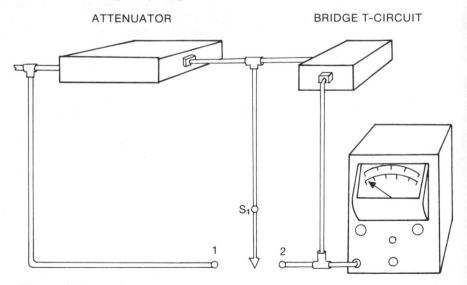

Figure 2-6: Harmonic analyzer

After you have the circuit constructed, all you need is an attenuator, preferably calibrated in dBs, and a power meter or VTVM. The equipment is set up as shown in Figure 2-6.

To make the measurement, first set switch S_1 to the 1 position. Now, assume that you read 3 milliwatts on the meter after adjusting the fundamental frequency suppressor network. The next step is to place the switch S_1 to the 2 position. We'll say that the new reading is 6 milliwatts (total power). Adjust the attenuator until you read exactly 3 milliwatts again. It should read 3 dB. This means that the harmonic power is 3 dB below the carrier power, i.e., one-half power. However, this is only a simple example and what you probably would want is something like 40 or 50 dB below the test frequency. If you don't have an attenuator calibrated in dBs, it isn't a big problem because you can make one quickly, as shown in Chapter 1.

Successful Dynamic Impedance Measurements

Have you ever looked at a piece of electronic equipment and wondered what the input impedance was? If so, here is a fast and simple way to put this trouble to rest forever. It is important that each piece of equipment under test has negligible reactance. In other words, it appears as a resistive load. There are two other important points: All connecting cables and components must have matching impedances when you set up this measurement and the measurement is for UHF TV and microwave equipment. Keeping this in mind, you'll have no trouble. There are only three steps necessary.

Step one: Connect the equipment as shown in Figure 2-7. Adjust your signal generator to a level 6 dB above the normal input to the equipment under test. This is to compensate for the series resistor in the circuit.

Step two: Using a VTVM, measure the voltages at E_1 and E_2. These are the output voltage of the generator and the input voltage to the piece of equipment under test.

Step three: Compute the ratio E_1 / E_2 and use the formula input impedance equals the value of R_1 divided by the voltage ratio minus 1. Or,

$$Z = R_1 / (V_R - 1)$$

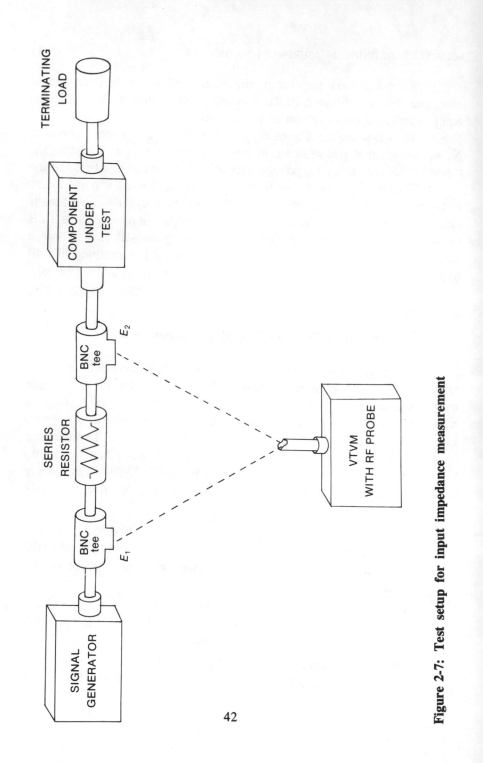

TERMINATING LOAD

COMPONENT UNDER TEST

E_2

BNC tee

SERIES RESISTOR

BNC tee

E_1

VTVM WITH RF PROBE

SIGNAL GENERATOR

42

Figure 2-7: Test setup for input impedance measurement

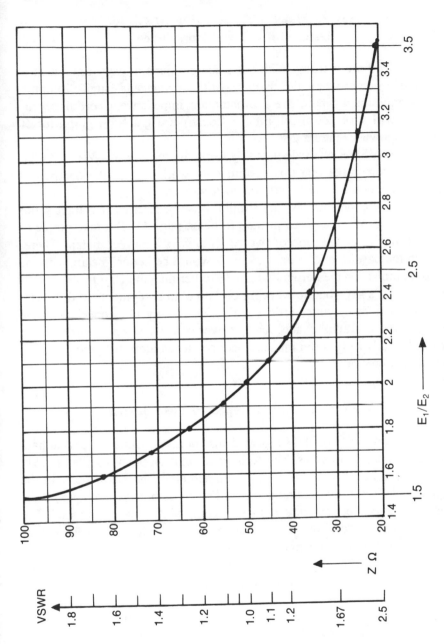

Figure 2-8: Graph for determining input impedance

How to Make a Time-Saving Graph for
Determining Input Impedance

If you are working with the same equipment every day, such as television systems where the characteristic impedance is usually 75 ohms, or test equipment where the characteristic impedance is ordinarily 50 ohms, it will save you much time to construct a graph. A graph for 50 ohms is shown in Figure 2-8.

The VSWR was calculated by using the formula $VSWR = Z_2/Z_1$. However, remember that this system only works where the component under test appears to have little or no reactance.

Making the graph is very simple. Assume a series of voltage ratios and use the procedure given in Step three on page 41. But take notice of the fact that a voltage ratio of two (2) is always a perfect match. Therefore, in the case of a 75 ohm system, it would be better to start with the bottom number on the vertical with a 40 rather than the 20. Or, as another example, let's say you are working on 600 ohm equipment. In this case, start the bottom of your vertical with 200. At any rate, once you have your graph (or graphs) made, hang it over your workbench and the next time you need to find the input impedance to an amplifier, etc., the job becomes a breeze.

What to Do About Repeated Fuse Burnout

A common trouble is repeated fuse burnout. Generally, what happens is something like this: The technician finds an open fuse; he quickly replaces the fuse. If the fuse blows immediately, he has a short in the equipment. If not, everything is okay. The equipment is sent back to the customer and the next day comes a complaint. The complaint? The equipment isn't working. The next scene is the service truck back out on the way for the same customer pickup.

As you can see from this scenario, repeated fuse burnout is not only aggravating but it can be costly. To be sure this problem doesn't plague you, it is only necessary to measure the amount of current being drawn through the fuse. For example, suppose that you find a hi-fi rated at 2 amps, with a blown fuse. Then you take a reading and find that the unit is actually drawing about 2.9 amps. Now, you immediately know that there is a partial short and there is an excellent chance that the fuse will blow again in a short time. It's apparent that this trouble is a short and not the fuse.

Or, let's say that a circuit breaker keeps popping out each time you reset it. You see that it is rated at 2 amps and yet you only measure 1 amp when you take a current measurement. Obviously you must replace the circuit breaker in this case. Remember, when in doubt, measure the current. It might save you a lot of time and effort.

Updated Semiconductor Techniques

Many technicians have a junk box full of solid state components under their workbenches. This chapter contains updated and successful ways to use common shop instruments to check these components. Furthermore, it will get these items back into service and reduce your operating cost at the same time. Also included are many simple but effective in-circuit troubleshooting tricks that may save you the painful job of de-soldering and pulling out a good transistor.

If you want to determine the overall high frequency response of two or more transistor amplifiers with nonidentical frequency response in cascade, it usually calls for fairly expensive equipment. But, it's also possible to do this job by drawing lines across some scales. The problem of nonidentical stages is especially important to anyone working with transistors because each stage is likely to have a different frequency response. This chapter also solves this problem.

Practical Transistor Measurements

One of the most destructive weapons in the shop can be the VOM (also the VTVM) in checking solid state components—unless you are lucky enough to have a Simpson 260-6XL or a Weston 670, which both have low power ohms capabilities. Figure 3-1 illustrates how a technician might get into trouble checking a transistor using a standard VOM ohms setting.

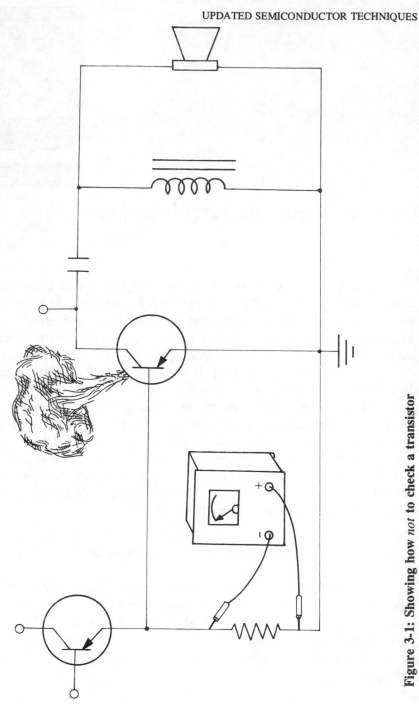

Figure 3-1: Showing how *not* to check a transistor

Notice that the ohmmeter is applying forward bias to the transistor. Now, considering the fact that many of these meters supply a short circuit current of 100 mA on the R × 1 range, you can see the reason for the smoke. This won't happen every time you use the R × 1 range but believe me, if someone is looking over your shoulder, it probably will! Yet, if you follow some simple rules, the VOM and its improved offspring, the VTVM, are well-behaved, low cost, easy-to-use, devices and good for a wide variety of transistor checks. For example, a simple breadboard circuit that can be used with your VOM to check leakage —also called *reverse current*—and to give you a rough idea of the beta of an unknown transistor is shown in Figure 3-2.

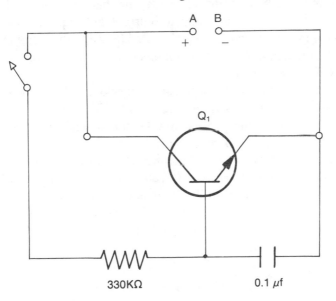

Figure 3-2: Low cost transistor tester

The transistor Q_1 is under test. There are two ways to determine the quality of the transistor. You can measure the collector-emitter reverse current, or *easier*, measure the resistance. The best transistor in a group will have the smallest leakage current, i.e., a large resistance.

To make the check, connect an ohmmeter to the terminals marked A and B, observing the correct polarities. Although if the transistor is a

PNP type, you will have to reverse the leads. Next, set the VOM to a high resistance range (this will produce the maximum voltage on the ohmmeter leads) and have the switch open. A good silicon transistor will produce very little, if any, noticeable downward deflection. This is because the reverse current is usually below 1 μA. But, if during this check you read zero resistance, junk it—the transistor collector circuit is shorted.

To check the transistor in the emitter-base circuit, set the range switch to R $\times$ 100 and close the switch. If the meter reads some lower value of resistance, the transistor is probably okay. Also, note the difference between the two readings because this difference will give you a rough idea of the beta of the transistor. For a more accurate value of beta, take a current reading in the collector and base circuits and use the formula $\beta = $ Ic / Ib. Ic is the collector current and Ib is the base current. The alpha of the transistor can be calculated by using the formula $\alpha = \beta /$ $(\beta + 1)$. Incidentally, if you should happen to know the α of a transistor, the beta will equal $\alpha / (\alpha - 1)$.

Semiconductor diodes also can be checked with a standard VOM. In most cases, it is best to switch to R $\times$ 100. This is because on a higher range the voltage on the probes may be too much and a lower range probably will produce too much current. Measure the resistance both ways and if you observe two different readings, one relatively high and the other low, the diode is good. If you know the actual polarity of the VOM leads, it is easy to identify the anode and cathode of the diode. A low resistance indicates that the positive lead is connected to the anode of the diode. On the other hand, a high resistance indicates that the positive lead is connected to the cathode.

How to Check Power Transistors

To make a check of the quality of a power transistor is just as easy. When checking the reverse current in the collector-emitter circuit of a silicon power transistor (in power transistors, this current can be up to 100 μA or so), set the VOM range to either R $\times$ 10 k or R $\times$ 100 k ohms. With the germanium type, it's better to drop down to the R $\times$ 1 k ohm range. One thing on your side when troubleshooting power transistors is that generally when a power transistor fails, it really fails: a complete open or a dead short, with the short being the most common trouble.

Bear in mind that power transistors normally use their metal cases for the collector lead. Furthermore, they are isolated from the chassis and, therefore, need silicon grease on both sides of the insulating material to insure good heat conduction to the chassis. After the transistor is mounted, it is a good idea to make a continuity check between the transistor case and chassis to be sure you have the proper insulation.

Successful Testing of Zener Diodes

Another solid state device that shows up on the workbench in an unknown condition is the zener diode. Here is a simple way to check its performance. Place a 22 k ohm (1 watt) resistor in series with the zener diode and connect this circuit to the ouput of your DC power supply. Next, connect a DC voltmeter across the zener as shown in Figure 3-3.

The zener will not start operating until you reach about 20% of its

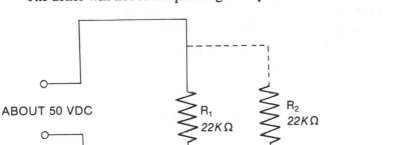
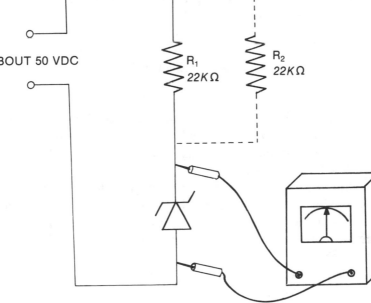

Figure 3-3: Checking a zener diode

maximum load. Therefore, this setup will require about 50 volts output from the DC power supply. To check the operation of the Zener, momentarily parallel another 22 k ohm (1 watt) resistor across the original resistor. If the zener is regulating properly, there won't be any change in the reading of the voltmeter.

Working with the UJT

It is also easy to make a simple go or no go check of a unijunction transistor (UJT) with a conventional VOM. To do this job, set the multimeter to the R × 100 range. Next, if a measurement is made between the emitter and base-1, forward conduction is indicated by a mid-scale reading. Reversing the test leads should produce a reading near infinity. When you check the resistance between base-1 and base-2 (with nothing connected to the emitter), the reading may be anywhere between several thousand ohms up to about 10 k ohms. However, exact value isn't important. What is important is that you get the same reading when you reverse the leads. Figure 3-4 illustrates each condition you should find during all checks if the UJT is okay.

A Guide to Checking SCR's

We started out saying that most VOM's can deliver quite a few milliamps on the R × 1 range. If yours does, you're in luck because this is just what we need to check a silicon controlled rectifier (SCR). The reason you need this current is to insure that there is sufficient "holding" current for the SCR under test.

Start by checking the resistance between the anode and cathode of the SCR with the gate left open. If the reading is low—regardless of polarity—into the trash can it goes because this is an indication of a short. Figure 3-5 shows how to check the triggering of an SCR with a small gate spike voltage.

Connect the positive test lead to the anode and the negative one to the cathode with the VOM set in the R × 1 range. Next, close the switch. This will cause a gate spike that should trigger the SCR and produce a low resistance reading on the VOM. This means the SCR is working and

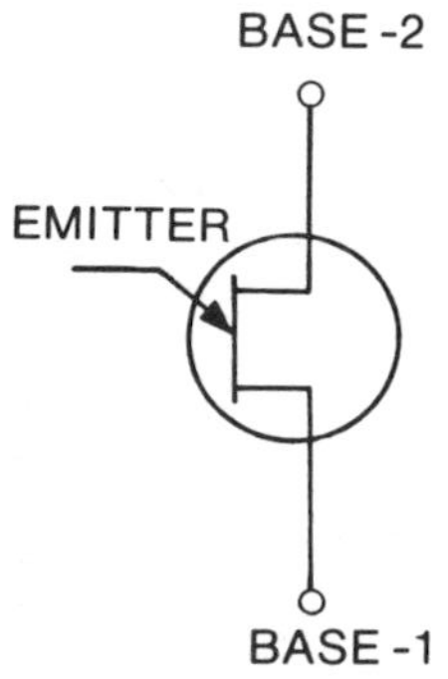

EMITTER	BASE-1	BASE-2	RESISTANCE READING
+	−		LOW
−	+		HIGH
−		+	HIGH
+	−		LOW
	+ −	+ −	SAME READING

BASE-2 ○———[EMITTER]———○ BASE-1

Figure 3-4: Resistance readings for a good UJT

53

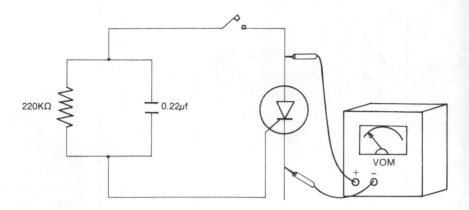

Figure 3-5: Checking an SCR

everything is fine. I have seen SCR's that won't trigger during the test. Should this happen, try placing about 1.5 volts in series with the RC circuit. This should do the trick. If it doesn't, discard it.

How to Test Triacs

Keep the VOM in the R × 1 range and it's simple to check any triac that you may have lying around. To begin with, a good triac will have a very high resistance in both directions if the gate is left floating. But to check it, place a resistor of about 10 or 15 ohms in series with the VOM and place the leads between either anode and the gate. One-half of the triac should conduct, depending on which anode (1 or 2) you connect

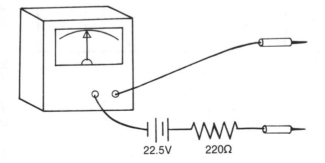

Figure 3-6: Test setup for testing triacs

to. To check the other half, reverse the ohmmeter leads and the ohmmeter again should show a low resistance.

There are some triacs that have a built-in diac gating element and it is necessary to use a larger voltage to check these. A 22.5 volt battery or a DC power supply of the same voltage will do. Switch your VOM range switch, in this case, to the 100 mA range and insert a 220 ohm resistor in series with the meter and power supply. Then the test procedure is the same as for checking SCRs. By the way, polarity is not important in this check. Figure 3-6 shows the test setup.

In-Circuit Troubleshooting

Transistors usually are soldered into the circuit and everyone soon learns to make every check possible before resorting to pulling the transistor. The following checks can save a lot of time, effort, and may improve your financial condition. An example of a transistor circuit with the voltage check points for troubleshooting is shown in Figure 3-7.

A point to notice is that the current through the emitter resistor R_3 is I_b plus I_c and therefore, is a good indicator of total current flow through the transistor and, as you can see, is easy to calculate. Collector current can be calculated in basically the same manner, i.e., $I_c = EM_3 / R_4$. Let's assume the voltage drop on M_2 or M_3 is high. Our first thought is that the transistor is defective. One simple way of determining if the transistor is at fault is to change the biasing voltage and then note the change of the voltage drop on M_2 or M_3. One common way of removing the emitter-base (forward) bias is to short the emitter to the base with a screwdriver.

If you short the emitter-base and the meter M_2 does not decrease in voltage reading, the transistor probably is in need of replacement. If the reading of the voltmeter does drop, the transistor is probably good.

Now, let's assume the voltage across M_2 or M_3 is abnormally low. In this case, place a larger plus voltage on the base, i.e., place more forward bias on the emitter-base circuit. A simple way to do this is to place a jumper wire between the collector and base. However, the positive collector voltage normally is too great so we temporarily place a resistor and jumper wire as shown in Figure 3-8.

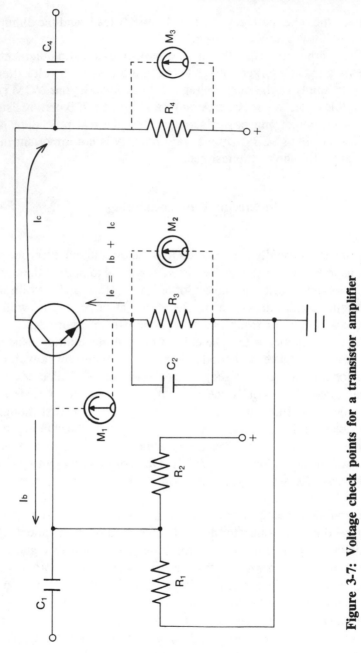

Figure 3-7: Voltage check points for a transistor amplifier

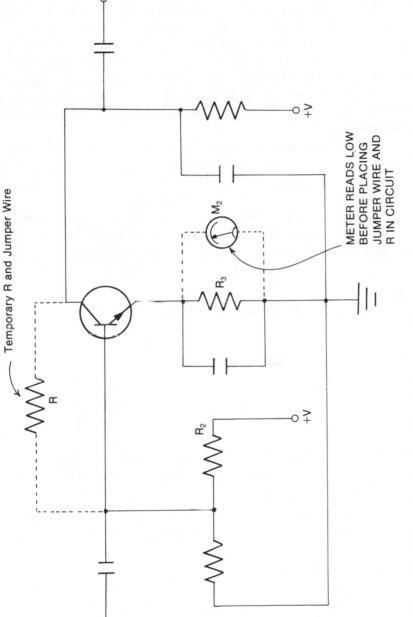

Figure 3-8: How to check a transistor by increasing bias

The value of the resistor is approximately the same as the value of the voltage divider resistor R_2. If after forward bias is increased, you read a higher voltage across M_2, the transistor should be good and in this case, you should check the voltage divider R_1, R_2. It is very posible that if C_1 shorts, R_2 will be burned. In properly operating transistor or tube amplifiers, the collector voltage or plate voltage almost always will be lower than the supply voltage. Therefore, this also is a good, quick check for in-circuit testing.

Another problem that frequently pops up is that an oscillator just will not start. Now, the first thing that comes to mind is that you don't have enough positive feedback, which is probably true. But it is possible to beat this problem by remembering the old phrase, "Don't raise the river, lower the bridge." How do we do this? Easy. Don't raise the feedback, lower the bias. In other words, if your oscillator won't start, try changing the bias about ± 20% before you start unsoldering that transistor.

Practical Troubleshooting Rules for Solid-State

Listed in the following section are a few more points that you should remember about solid-state equipment.

1. Be careful when using a standard VOM because it can permanently damage the small electrolitic capacitors used in much of today's equipment. Many are rated at a working voltage of only 3 volts.

2. Another point: Your oscilloscope should have a deflection sensitivity of 10 millivolts per inch or better to work with modern solid-state components.

3. Next, a blocking capacitor should be placed in series with any signal generator test lead before injecting a signal into the circuit under test.

4. To check a PC board, use the eraser end of a pencil and slightly press the board. If the signal returns, you have a broken connection or the board is cracked. This won't work every time, but it will in many cases.

5. Finally, when using mercury cells, be sure and note that they have opposite polarities to the penlite cells.

Using a Ruler to Find Amplifier Frequency Response

Many times it is necessary to connect two, three, or more amplifiers in series. Anytime this is done, it causes the overall frequency response to decrease to less than that of the narrowest stage. The overall frequency response of identical RC coupled stages in cascade can be calculated by using the formula $F = \sqrt{2^{1/n} - 1}$ where n equals the number of identical stages and F is a number that can be used to calculate the shrinkage factor. For example, two stages will produce an F of 0.644. Now, multiply this number by the high frequency response of the worst case amplifier and you have the overall frequency response of the two. The following table can be used to find the required F factor for up to eleven amplifiers in cascade.

NUMBER OF STAGES	F FACTOR	NUMBER OF STAGES	F FACTOR
2	0.6435	7	0.3256
3	0.5098	8	0.3008
4	0.4349	9	0.2829
5	0.3856	10	0.2679
6	0.3499	11	0.2550

Table 3-1: High frequency reduction factor for identical stages

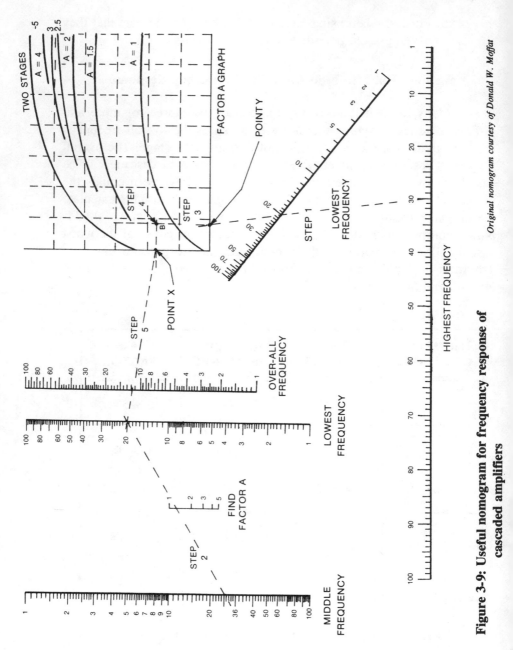

Original nomogram courtesy of Donald W. Moffat

Figure 3-9: Useful nomogram for frequency response of cascaded amplifiers

As you can see, as long as all stages are identical it is a simple procedure to determine the overall high frequency response. However, it has been and is a tough problem if the stages involved are not identical and you don't have a sweep generator. The accompanying nomogram reduces the job to using a ruler as a straight edge. See Figure 3-9.

First, notice that the frequency scales are conveniently set up so that all you have to do is add any number of zeros to utilize the nomogram in any range of frequencies. For example, if you are working in the region of 1 MHz, you let the 10 in the middle of each frequency scale stand for 1 MHz. This makes the ends of each scale 100 kHz and 10 MHz. Be sure that you add zeros to *all* frequency scales when setting up for different parts of the frequency spectrum.

Let's assume that you have three transistor amplifiers in cascade and their individual frequency responses are 20, 25, and 30 kilo hertz. Now here is what you do to find the overall frequency response, step by step.

Step 1. Find 30 on the highest frequency scale and 20 on the lowest frequency scale. Place a straight edge across these two points and draw a line as shown by Step 1 in Figure 3-9. Extend the line so that it passes through the upper graph as shown.

Step 2. Look at the line labeled Step 2. Notice that we have drawn a line through 25 (middle frequency scale) and 20 on the lowest frequency scale. Now, notice that this line passes through the Find Factor A scale. The reading on the Find Factor A scale tells you which curve on the Factor A graph to use.

Step 3. Now, go back to the Factor A graph and draw a line straight up from Y to B. Point Y is where Step 1 line crosses the horizontal axis of the graph. Point B was estimated because the Factor A is between 1 and 1.5 (about 1 1/4).

Step 4. Draw a line straight out from point B to the vertical axis (point X).

Step 5. Finally, draw a line from point X to 20 on the lowest frequency scale. Read the number on the overall frequency scale that this line passes through and you have the response of your three stages. In this case, it is about 12 kHz.

If you have two transistor amplifiers that have different frequency responses, it is simpler because there is no factor A involved. Here's how to do this one step by step.

Step 1. Draw a line as before, from the highest frequency scale up through the lowest frequency scale to the bottom of the Factor A graph.

Step 2. Draw the line straight up and locate point B on the two-stage curve.

Step 3. Draw a line straight out from point B to some point X that will be higher up the graph.

Step 4. Draw a line from your point to the lowest frequency scale. Read where this line crosses the overall frequency scale and this is your overall frequency response, and that's all there is to it.

A Guide to Troubleshooting IC's

Are you confused by operational amplifier integrated circuits (**OP AMPS**)? For instance, how do you check one of the beasts? This chapter shows how to construct a simple **OP AMP** tester that can handle most of the internally compensated or externally compensated types. Or, maybe the problem is how to troubleshoot a complementary-symmetry-metal-oxide semiconductor **IC** (**CMOS IC**). If you don't understand these **IC**'s you should, for although you may not want to specialize in integrated circuits the day is coming, and soon, when a working knowledge of **IC**'s can make the difference in your technical survival.

Every technician has run into the problem of translating the hieroglyphics used in talking about integrated circuits—for example, **DTL, TTL, CMOS,** and **MOSFET**. There is an easy-to-use, practical glossary at the end of this chapter that will straighten out some of the kinks you may have in this area.

Practical OP AMP Techniques

By now, everyone should be more or less familiar with the name operational amplifier (**OP AMP**) and that its name came from the theory of feedback amplifiers. In fact, try to use one without the proper feedback and it will refuse to budge no matter what you do. Figure 4-1 graphically illustrates the problem. The gain of the amplifier is designated (Af), (A) is some fixed value, and Rf is a fixed value of feedback resistor.

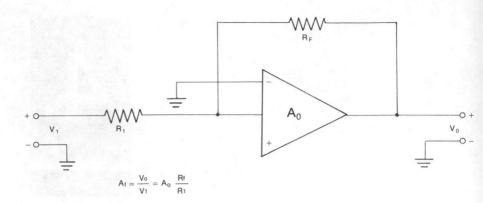

$$A_f = \frac{V_o}{V_1} = A_o \frac{R_f}{R_1}$$

Figure 4-1: Basic OP AMP circuit—Voltage gain depends on the amount of feedback

How much feedback is proper? It depends. Any operational amplifier loses gain as frequency is increased and, therefore, works better in the lower part of its bandwidth. The trick is to produce enough excess gain to keep the circuit working properly at the highest frequency you want to work with. To be on the safe side, adjust your feedback for a minimum excess gain of ten times what the circuit needs at the frequency of interest. Using this ratio won't give the best performance in the world (you will have between 5 and 10% operational error), but it will produce the best overall gain for the widest band of frequencies.

If you are not familiar with the symbols used when discussing **OP AMP**'s, it's possible that you may get into trouble. Figure 4-2 shows a schematic drawing and some common pin arrangements.

The inverting input is normally shown as a $(-)$ and the noninverting, as a $(+)$. However, watch it! In the case of **OP AMP**'s, the inputs marked $(-)$ and $(+)$ have nothing to do with the supply voltage. When speaking of the supply voltages, they are called $V+$ and $V-$. But, again be careful because the power supply used with **IC**'s is a bi-polar supply. In other words, it has a negative and positive voltage output with respect to ground.

Now, back to feedback. Here's a general rule to use when troubleshooting **OP AMP**'s. If the feedback goes from the output to the noninverting input $(+)$, you should have a logic type output. Or conversely, if the feedback goes from the output to the inverting input $(-)$, you should have a linear output. In this case, gain $= R_f / R_1 + 1$.

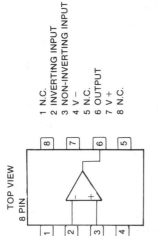

TOP VIEW
8 PIN

1 N.C.
2 INVERTING INPUT
3 NON-INVERTING INPUT
4 V −
5 N.C.
6 OUTPUT
7 V +
8 N.C.

TOP VIEW

1 N.C.
2 INVERTING INPUT
3 NON-INVERTING INPUT
4 V −
5 N.C.
6 OUTPUT
7 V +
8 N.C.

Pin arrangement courtesy of Signetics, Data book © 1974, 81
E. Arques, Sunnyvale, Ca.

SCHEMATIC SYMBOL

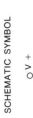

V +

OUTPUT

INVERTING INPUT

NON-INVERTING INPUT

V −

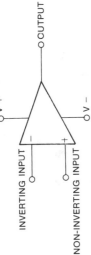

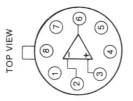

TOP VIEW
14 PIN

N.C. 1   14 N.C.
COMP 1 2   13 N.C.
*GUARD 3   12 COMP. 2
INPUT 4   11 V +
INPUT 5   10 OUTPUT
*GUARD 6   9 N.C.
V − 7   8 N.C.

*UNUSED PIN (NO INTERNAL CONNECTION)
TO ALLOW FOR INPUT ANTI-LEAKAGE

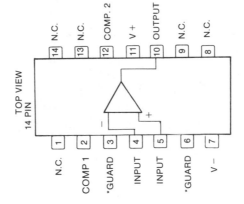

Figure 4-2: Pin arrangements of operational amplifiers and schematic symbol

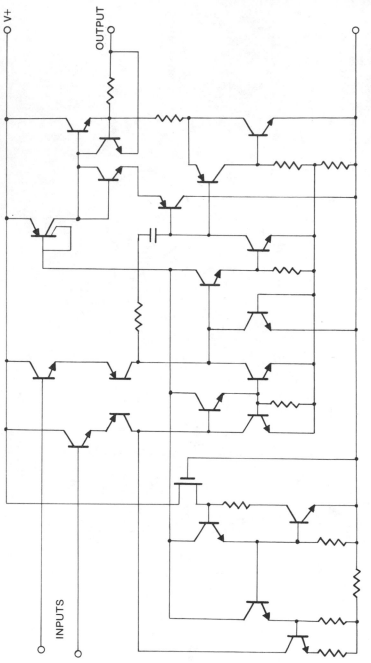

Circuitry courtesy of Signetics, Data Book
© 1974, 811 E. Arques, Sunnyvale, Ca.

Figure 4-3: OP AMP equivalent circuit

The next question. Is there a bias current to the inverted and noninverted inputs? Well, put it this way. If it isn't provided, the circuit won't work. During troubleshooting, you will find this bias current provided in different ways. For example, through resistors or coils to ground, from the output back to the input, or in some cases, from another reference voltage. Typically, these bias currents are about 100 nano amps. Because of this tough requirement, you probably will find a **CMOS** array (explained later in this chapter) preceding many **OP AMP**'s. Due to the very low bias current, the device is frequently called a zero input **OP AMP.** Figure 4-3 shows a typical **OP AMP** equivalent schematic.

How to Construct and Use a Simple OP AMP Tester

As has been explained, any **OP AMP** requires a bi-polar power

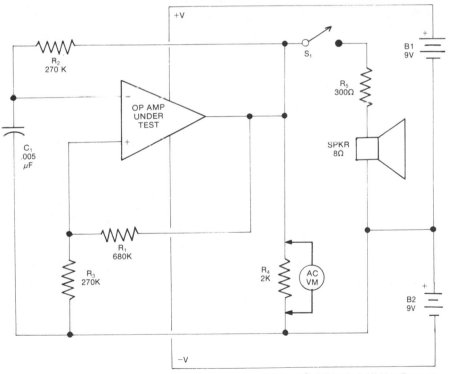

Copyright © Ziff-Davis Publishing Company
Reprinted by permission of Popular Electronics Magazine

Figure 4-4: OP AMP tester

supply. When using the following tester, it is easier to use a full-wave zener regulated power supply with a negative and positive voltage output with respect to ground, if you have one. However, two batteries of the proper voltage can be used (a 9-volt transistor radio battery for the familiar 741 **OP AMP** is an example) but you probably will find that they will run down rather quickly. I have found that unless you do a lot of work where portability is important, the power supply is better and cheaper in the long run. Figure 4-4 shows the circuit diagram for the tester.

This tester is no more than a simple, free-running multivibrator and the time constant R_1C_1 sets the frequency. To check the quality of an **OP AMP,** place an oscilloscope's leads across R_4. It should read very near 12 volts peak-to-peak with R_1 set at approximately midscale if you are checking a 741 or anything similar. Or if you wish, place a 330 ohm resistor in series with an 8 ohm speaker and you can hear the square wave output by paralleling the speaker and resistor with R_4. The voltage across R_4 should drop to about 6 volts peak-to-peak with the speaker in the circuit. Sometimes a speaker really helps because you don't have to try and keep your eyes on both the test probes and read the meter at the same time.

Troubleshooting Tips for IC's

If you have ever soldered a 16 pin **IC** directly to a **PC** board and found that some of the sections were not functional, don't feel bad. You're in good company. Take it from me . . . don't trust anyone. Test them yourself to be sure. Also, be very careful to make sure that the IC is properly oriented before soldering. In fact, it is much better, and safer, to use sockets and leave the **IC**'s out of the circuit during changes if it is at all practical.

Like most solid state devices, **IC**'s are sensitive to abuse. For example, they are *easily* destroyed by using too large a soldering iron. Generally, it is best to use a low wattage (25-35 watts) soldering iron when mounting them. But, there is more than one way to lose an **IC**. Insert it or remove it from a circuit with power on and it's gone!

Another fact of life about **MOSFET**'s and **CMOS IC**'s is that they must be kept in conductive foam (which usually is used to pack **MOS IC**'s) or, somehow the leads must be shorted. One good way to do this is to wrap aluminum foil around the leads. *Note: Never put* **MOS IC**'s *in styrofoam or snow.*

Some Important Points About Grounding
When Troubleshooting IC's

Everyone who has worked with microwave equipment knows that you must ground your body while handling crystal diodes because of static electricity. The same rule holds true when working with **MOS IC**'s. This is especially true of microcomputer chips and their memory chips. It's best to place a grounded metal plate on the workbench and keep your arms in contact with the chassis. Not only must you ground yourself, but you should use grounded soldering tips and grounded test fixtures as well. Always check the test leads of each piece of test equipment as well as the soldering iron tip for a voltage between the tip and ground. And, do this check twice by *reversing* the AC plug.

How to Work with CMOS IC's

The metal oxide semiconductor (**MOS**) technique has proved very good for making large scale integration (**LSI**) chips because it is possible to package many circuits on the same chip. But we never get anything without a trade off. Therefore, **MOS** chips are slower operating than the familiar transistor-transistor logic (**TTL**) type. Before we jump to any conclusions, it should be mentioned that **TTL** isn't good for **LSI** because they generate too much heat and use up more space. For these reasons, **TTL**'s are usually found in medium scale integration (**MSI**) and small scale integration (**SSI**) devices.

Although **CMO**'s are not as fast as **TTL**'s, it is a way to make **MOS** chips that use almost no power and therefore, is great for uses such as electronic watches or other portable equipment where you must depend on batteries for power.

What should you see on the output of a **CMOS IC**? It depends on the input and the circuit connections. For example, a **CMOS** stage will have a balanced output when the input signal voltage is one-half the source voltage V_{dd} if it is used in logic circuits. In case you're not familiar with **MOS** language, V_{dd} is the positive supply voltage (old B +) and V_{ss} is the negative supply voltage (normally grounded).

By a balanced output, we mean that the output is halfway between V_{dd} and V_{ss}. If the input level is increased or decreased, it will cause the stage to flip the output to a high or low level. A basic **CMOS** stage is shown in Figure 4-5.

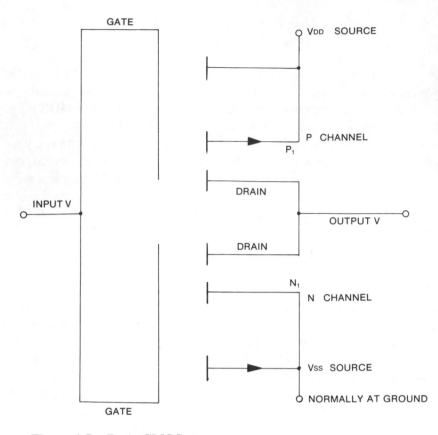

Figure 4-5:   Basic CMOS stage

Notice, there are two **MOSFET**'s that are effectively two equal value resistors. Therefore, the output is one-half Vdd and it is balanced. Now, suppose you increase the input voltage, approaching Vdd. This produces less bias on P_1 and N_1, which unbalances the voltage divider, resulting in a drop on the output. And if you decrease the input voltage, you get a rise on the output. Surely, an electronic device of any kind can't be easier to check than this one. However, watch those unwanted pulses (such as noise) because they may trigger the circuit without your help. An **IC** digital logic memory probe can be used in establishing the presence of these unwanted signals.

What You Need to Know About CMOS IC's

As always, there are a few more rules of the game that must be followed. One of these rules pertains to the input signal voltage limits. It is absolutely necessary that you *never* use an input voltage level greater than the supply voltage, V_{dd}. A safe input voltage is always less than (or equal to) V_{dd} and more than (or equal to) V_{ss}, and if this rule is not followed, the chip probably will be destroyed. An equally important point: *Never* apply an input signal without power on. Without power on, you won't have a balanced condition and this can result in a burned out section of your **IC**.

How to Prevent Troubles When Working with PC Boards That Have CMOS IC's

When working with **PC** boards that have **CMOS IC**'s, it is necessary to terminate all leads with a 1 megohm resistor so the leads won't be left floating when the **PC** board is removed from the chassis. Also, don't allow any of the **CMOS** inputs to remain open. Tie them in parallel with the input from the same gate. By the way, you want to save these unused sections because you can always connect to them if any of the operating sections become defective.

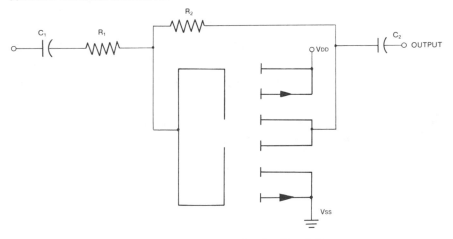

Figure 4-6:   CMOS inverter with negative feedback for linear operation

Using a CMOS IC in Linear Applications

Like **OP AMP**'s, if a **CMOS** is used in linear applications, it is best to use negative feedback. If we take Figure 4-5 and connect the components as shown in Figure 4-6, we will have a linear amplifier.

The gain of the stage is set by the ratio of the values of R_2 to R_1. As you can see, starting with a megohm for R_1 and R_2 will produce a gain of unity. However, about the best you can do is a gain of 10 if distortion is to be kept to a minimum (less than 1%). For higher gains, just cascade each stage in the **IC** and the overall gain will be the product of each stage gain. Incidentally, the input impedance of the stage is set by the value chosen for the resistor R_1, and the capacitors C_1 and C_2 are simply coupling capacitors. Figure 4-7 shows some typical **MOS** packages.

Important Rules for Handling and Troubleshooting IC's

Here are a few important points that you want to remember during handling and troubleshooting of **MOS IC**'s.

1. It is possible to permanently damage an **MOS IC** merely by leaving it on the workbench without some type of shorting device. However, the newer **MOS IC**'s have been improved and the problem has been reduced (but not eliminated).

2. *Never* wear silk garments or any other clothing that may develop a static charge (such as wool) when working on **MOS IC**'s.

3. *Never* leave the input pins of a **MOS IC** unterminated on a **PC** board.

4. Do all soldering with an isolated soldering iron and be sure that all test equipment is properly grounded, as described earlier in this chapter.

5. *Do not* store **MOS IC**'s in plastic containers. Also, always store or ship a **MOS IC** with all leads shorted or in conducting foam.

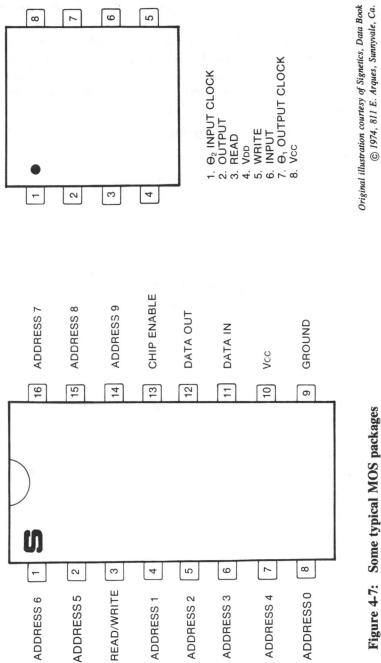

1. Θ₂ INPUT CLOCK
2. OUTPUT
3. READ
4. V_DD
5. WRITE
6. INPUT
7. Θ₁ OUTPUT CLOCK
8. V_CC

Original illustration courtesy of Signetics, Data Book
© 1974, 811 E. Arques, Sunnyvale, Ca.

16. ADDRESS 7
15. ADDRESS 8
14. ADDRESS 9
13. CHIP ENABLE
12. DATA OUT
11. DATA IN
10. V_CC
9. GROUND

1. ADDRESS 6
2. ADDRESS 5
3. READ/WRITE
4. ADDRESS 1
5. ADDRESS 2
6. ADDRESS 3
7. ADDRESS 4
8. ADDRESS 0

Figure 4-7: Some typical MOS packages

A List of Solid-State Abbreviations Frequently Needed in the Shop

When you pick up a trade journal or manufacturer's instruction sheet, it sometimes seems as if the author has written in code. And, in fact, he has. The names of integrated circuits have become so long that technicians who use them all the time use abbreviations. The following is a key to the code:

a-d	analog to digital
CCD	charge-coupled devices
CDI	collector diffusion isolation
CMOS	complementary-metal-oxide semiconductor
DAA	data access arrangement
DCTL	direct-coupled transistor logic
DIP	dual in-line packages
DMM	digital multimeter
DTL	diode transistor logic
DVM	digital voltmeter
ECL	emitter coupled logic
FET	field effect transistor
IC	integrated circuit
IGFET	insulated gate field effect transistor
I^2L	integrated injection logic
IMM	integrated magnetic memory
J-FET	junction field effect transistor
LED	light emitting diode
LSI	large scale integration
MOS	metal-oxide semiconductor
MOSFET	metal-oxide semiconductor field effect transistor
MSI	medium scale integration

OMS	ovonic memory switch
OP AMP	operational amplifier
OTS	ovonic threshold switch
PIN	P-intrinsic-N Material
PMOS	P-channel metal oxide semiconductor
PROM	programmable read-only memory
RAM	random access memory
RCTL	resistor-capacitor transistor logic
ROM	read-only memory
RTL	resistor transistor logic
SATO	self-aligned thick oxide
SOS	silicon on sapphire
SSI	small scale integration
TTL	transistor-transistor logic
UART	universal asynchronous receiver/transmitter
V-ATE	vertical anisotropic etch
VIP	vcc-isolation with poly silicon backfill
VTR	video tape recorders

Troubleshooting Delta and
In-line Three-gun Color TV Receivers

In this chapter you'll find practical troubleshooting and alignment procedures that will be very helpful to you when working on color TV receivers. For example, a step-by-step explanation of the convergence adjustment for a delta and slotted mask in-line, three-gun color CRT is explained. Also how to troubleshoot the AGC circuit, quick and simple adjustments to eliminate adjacent channel sound and video interference, and how to check a modern horizontal oscillator are some of the practical checks and troubleshooting techniques that you will find in the following pages.

How to Make a Color TV Gray Scale Adjustment

Frequently the technician must determine if a three-gun shadow mask type color TV (also called a 3-gun delta type CRT) needs a gray scale adjustment. To do this, turn the color intensity control all the way off and ideally, you shouldn't be able to detect any color on the viewing screen. If you can detect color tints across the entire face of the CRT screen, it's an indication that the receiver needs a gray scale adjustment.

It should be pointed out that you can't get all the color out with a gray scale adjustment if the purity adjustment is off. What will happen in this case is that you'll see splotches of color on the screen even after the gray scale adjustment is completed if the purity is out of adjustment.

However, it's generally best to do the gray scale adjustment first and then finish up with the purity adjustment described in the next section.

On a delta type three-gun CRT (the single gun color TV picture tube doesn't have purity and convergence adjustments) you'll normally find three drive controls and three screen control adjustments. Figure 5-1 shows the typical layout of the controls found on the chassis apron.

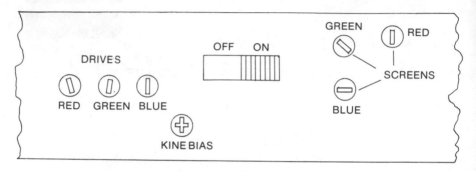

Figure 5-1: Service switch and controls used for a color TV gray scale adjustment

Your first step when making the adjustment is to kill the vertical sweep with the service switch. Next, turn the three drive controls all the way clockwise (all the way up) and turn the three screen controls all the way counter clockwise (all the way down). Now, you want to turn each of the screens back up and try to get a white as near perfect as possible. Sometimes one of the screens may not produce color. If this happens, turn the tube bias control up and it should do the trick.

Finally, place the service switch back to the full sweep position and examine the CRT screen for color tints. If you find any adjust the drives, which should bring in a satisfactory presentation.

A Practical Guide to Color Purity Adjustments

The objective of a color purity adjustment is to make each of the three electron beams in the CRT strike its correct color phosphor. If they don't, you will see patches of color in a black and white picture or bad splotches of color in a color picture. If you use a magnifying glass and look closely at the viewing screen, it's easy to see where the electron beams are off the correct color phosphor. Generally, if the beams are incorrect around the perimeter of the screen, it is an indication that the deflection yoke needs adjustment. Try moving it back and forth on the neck of the tube and it probably will clear up the problem. However, sometimes you may find splotches of color in the center of the screen. In this case, it's the purity ring, or possibly the CRT or its holder have become magnetized. Incidentally, if you're working with a toroid precision static (PST) type yoke, it cannot be moved because it is firmly connected to the neck of the tube. You'll find these systems discussed later in this chapter.

Once you get the hang of it, the purity adjustments are very simple. Start by turning off both the blue and green screens, and then to eliminate all residual magnetic fields, it is best to use an external degausing coil on the CRT. Next, while watching the center of the CRT viewing screen adjust the purity tabs for the best red display. See Figure 5-2 for purity ring magnets.

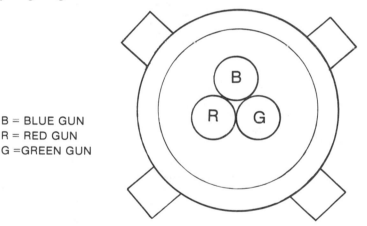

B = BLUE GUN
R = RED GUN
G =GREEN GUN

Figure 5-2:   Delta 3-gun type CRT purity ring magnets

Now adjust the deflection yoke, paying particular attention to the landing position of the beam around the outer area. In most cases you will have to do the adjustments several times to produce the best red possible. Also, more than likely you will have to redo the gray scale adjustment as described in the last section. Figure 5-3 shows the location of the purity rings, deflection yoke, and so on.

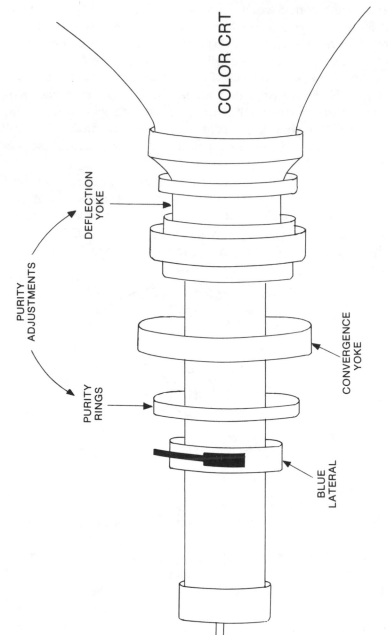

Figure 5-3: Location of purity adjustment components on a 3-gun shadow mask color picture tube

How to Make a Convergence Adjustment

In order to produce a sharp and clear color picture in a delta type 3-gun CRT, it is necessary that each electron beam strike only those color dots assigned to it. The purpose of the convergence adjustment is to correct any misalignment of the beams. Generally, this problem comes about as the receiver gets older or because the deflection yoke or convergence yoke has been replaced. The symptom is that colors will bleed and distort the picture. This is especially noticeable on a black and white picture.

The best way to start the convergence adjustments is to use a dot-bar generator and do the center of the screen first. The dot-bar generator is connected to the TV receiver antenna terminals with the set tuned to one of the low VHF channels and all equipment turned on. When you have the proper settings on the dot-bar generator you should see the entire TV screen filled with dots. You should find five sets of controls that control the center, top, bottom, left side, and right side and you must

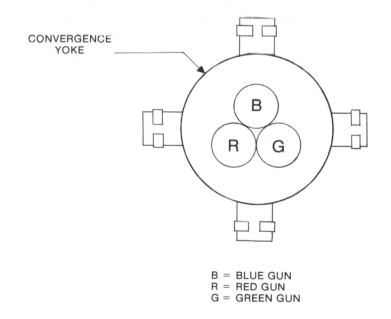

B = BLUE GUN
R = RED GUN
G = GREEN GUN

**Figure 5-4: Position of radial convergence magnets on
a standard delta type system**

watch each section as you make the adjustments. An old trick is to place a mirror in front of the TV to watch the viewing screen as you make these adjustments and what you want is to merge the red, green, and blue dots until they produce a single white dot of light.

As has been said, it's best to do the center of the screen first. There are four adjustments that consist of three static magnet adjustments and one lateral adjustment. The static adjustments are found on the convergence yoke: the red at the 8 o'clock position; green at 4 o'clock; and blue at 12 o'clock. See Figure 5-4. Usually, you'll find the blue lateral adjustment behind the convergence yoke. (See Figure 5-3.)

Key Steps to Adjusting a Color CRT's
Outside Perimeter for Convergence

The first thing to point out is that you won't find the following adjustments in all TV receivers. For example, in some cases the in-line gun system is complemented by a precision static toroid yoke that is permanently bonded to the neck of the CRT. Therefore, normally you replace yoke and tube together. Between the factory adjusted, in-line beams and permanently affixed yoke, (factory adjusted in place) all of the 12 dynamic convergence adjustments to be explained in the following discussion are eliminated. Figure 5-5 shows the in-line gun system compared to the standard delta system. How to make the convergence adjustments on a slotted mask color picture tube is explained after the following discussion.

Now, let's get back to the adjustments. After you have the center of the screen converged and assuming that you have to make the adjustments, your next step is to converge the top, bottom, left, and right side of the screen. The adjustment controls usually are placed on a convergence board. There are several different control arrangements; however, their functions are all basically the same. Figure 5-6 shows a typical layout of the 12 adjustments that you may encounter.

To start, it's best to use a cross-hatch pattern to make these adjustments. Your objective is to merge the red, green, and blue cross-hatch pattern so that you will see a perfect white light cross-hatch. A word of warning! If there is a high voltage problem, sweep section problem, or a bad CRT, it may not be possible to make a good convergence. Also, very

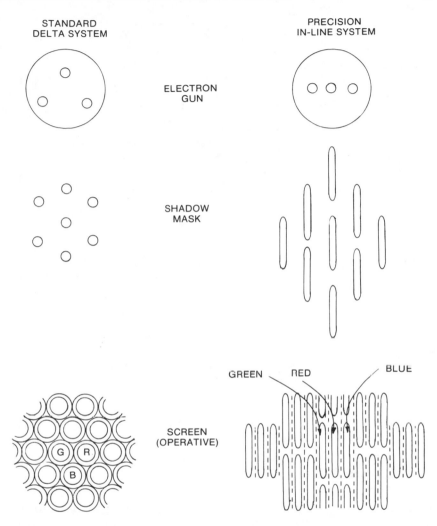

Figure 5-5: Delta 3-gun CRT in comparison to a precision in-line 3-gun CRT

frequently you won't be able to produce a perfect white cross-hatch. In this case, back off from the TV viewing screen and if it looks as if there isn't any color bleeding at about five or six feet, most viewers will be satisfied.

There are several adjustments called dynamic adjustments. There

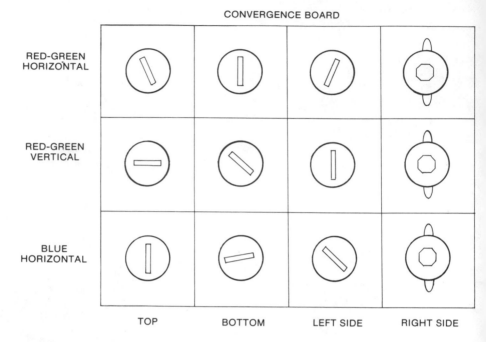

Figure 5-6: The 12 adjustments typically found on a delta type 3-gun color CRT convergence board

is one for the red-green horizontal bars, another for the red-green vertical bars, and one for the blue horizontal bars. Incidentally, you won't find one for blue vertical because this is the reference to which everything else is adjusted. See Figure 5-3.

    Referring to Figure 5-6, you will see that there are three adjustments for each section of the screen. Although you normally work on one section at a time, you'll find that there will be quite a bit of interaction between top and bottom and left and right sides. Because of this interaction, it's a good idea to use two hands so you can correct between the left side, right side, and so on. For example, when you are adjusting red-green horizontal left side, keep your other hand wtih the tuning tool on the red-green horizontal right side, and if the left side adjustment moves the right side, you can adjust it easily back in place. Many technicians use the following procedure to converge on the reference blue line:

Step 1. Adjust top and bottom red-green vertical.

Step 2. Adjust left side and right side red-green vertical.

Step 3. Adjust top and bottom red-green horizontal.

Step 4. Adjust left side and right side red-green horizontal.

Next, there are two more adjustments necessary for convergence on the red-green horizontal bars.

Step 5. Adjust top and bottom blue horizontal.

Step 6. Adjust left side and right side blue horizontal.

There are a few other points of which you should be aware. One is that if you are tuning one of the right side adjustments and all of a sudden get no further change, stop and rotate the adjustment (tuned coil) back to where you again see an effect. The reason is that when you see no further effect, you have passed the coil's resonant point, and any additional adjustment will produce nothing but wasted effort. Also, in almost every case you will have to do the adjustment several times to get the results you desire and you'll more than likely have to redo the gray scale adjustment described in the preceding section of this chapter.

Working on the In-line 3-gun Color Picture Tube

When you take the back off of some sets, you won't find a dynamic convergence board; no dynamic convergence yoke, and no blue lateral magnet on the neck of the tube; only one picture screen control. None of these parts are needed in the system introduced by RCA some time ago. Most of the adjustments have been done at the factory—even the purity and gray scale adjustments. With this system, there are no dynamic convergence adjustments, such as were just explained. You'll find the convergence adjustments mounted in a small assembly, just behind the deflection yoke. There isn't any convergence yoke. There are four rings in pairs. Two of them develop four-pole fields to move the blue and green beams equally in opposite directions. See Figure 5-7.

The other set develops six-pole fields and also moves the blue and green beams equally, but in the same direction. See Figure 5-8. What this boils down to is that all you have to do to converge the CRT is move the blue and green beams. The red is not moved and furthermore, the deflec-

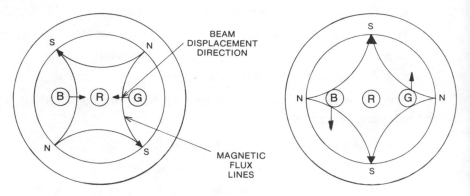

**Figure 5-7:   Two 4-pole magnets that move the blue and
green beams of an in-line color, 3-gun CRT**

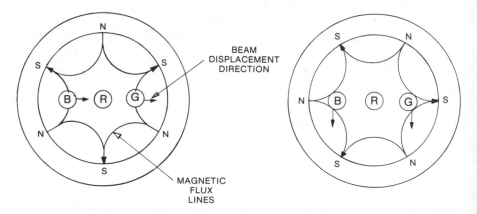

**Figure 5-8:   6-pole magnets that move the blue and
green beams of an in-line color, 3-gun CRT**

tion yoke isn't moved. In fact, you must use a heat gun to move it should
it become necessary to do so.

Troubleshooting the AGC Circuit

Generally, you should suspect that you have an automatic gain
control (AGC) problem when you have a loss of signal or distortion in the
video of only the strong stations. However, it's quite possible that you

may not have an AGC trouble because you could get the same symptoms with a bad IF amplifier or video detector. Therefore, it is necessary to make a simple test.

One way to do this is to use a bias box (a variable bench type power supply with a negative and positive voltage output). The test setup to check the AGC circuit is fairly simple. Attach the bias box to a resistor that is isolated above ground. You will find this circuit between the first IF amplifier and the keyed AGC amplifier. See Figure 5-9.

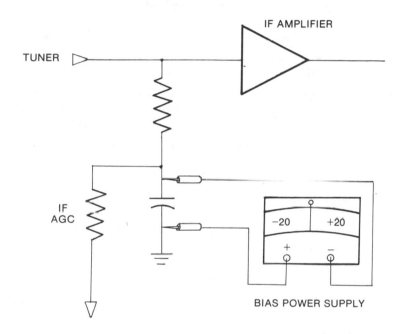

Figure 5-9: **Test setup for IF amplifier AGC trouble-**
shooting

After you have made the proper connections, turn on the TV and attempt to eliminate the problem by adjusting your bias box voltage. If you can, the AGC circuit is your problem. If not, it's somewhere else —possibly in the IF circuit or video detector.

When making AGC connections with a bias box, you should remember that you sometimes need negative bias and sometimes positive bias. For example, in the case of a forward biased common emitter NPN transistor circuit, the bias to the base must be negative. However, under the same circumstances a PNP requires a positive voltage on the base. Also, remember that with a common transistor emitter amplifier, zero voltage on the base will produce zero collector current. But in a common cathode tube amplifier, zero bias on the grid may cause considerable plate current to flow.

An Easy Way to Check AGC Problems at the RF Amplifier

An AGC problem in the RF amplifier will show up as snow on the strong channels and the weaker channels will appear good. Try disconnecting the AGC line from the RF amplifier. If the snow disappears, it's the AGC. But if you still have snow, (particularly if it's on all channels) check the RF amplifier because more than likely it is the problem. If you're working on a vacuum tube type set, your best bet is to substitute the RF amplifier tube as a first step. Ninety percent of the time this will clear up the problem.

Troubleshooting the Horizontal Oscillator

A very common circuit used for the horizontal oscillator in TV receivers is the cathode-coupled multivibrator. You will remember that the coupling capacitor and grid resistor are the frequency determining factors in a multivibrator. Therefore, all we have to do is short out the afc grid and it will place the oscillator in free-running condition. See Figure 5-10.

To check the oscillator, all you need to do is watch the picture and follow these steps:

Step 1. Ground the afc grid (See figure 5-10). This will place the oscillator in a free-running configuration.

Step 2. Place a jumper lead across the stabilizing coil. Warning: Don't ground it because, as you can see by the schematic, the coil is at B+.

Sometimes the oscillator will not run when you short the coil. If this happens, lift an end of resistor R (the one just to the right of the coil) and place a 15 k resistor in its place and it should do the trick. Incidentally, be sure to remove the 15 k resistor and resolder the original resistor when you're through with the check.

Now, to check the multivibrator, adjust the horizontal hold control until you see only one picture on the screen. If you can get the picture to hold stationary for even a second or so and the sides are straight, your problem is in some other circuit. What you are checking is to see if the free-running frequency of the oscillator is close enough to the right frequency to enable the afc to lock in.

If you can't get the oscillator to operate as described, it's an indication that there is a bad component. If it's impossible to produce the correct frequency, it's probably a faulty coupling capacitor or grid resistor. When you check the grid resistor, be sure and note that the horizontal hold adjustment is also part of the circuit that includes the resistor marked R_2 in Figure 5-10.

If the oscillator does run okay, take the jumper lead off the stabilizing coil. When you do this, it's possible that the picture will not hold. If it won't, tune the coil until you have the picture locked in. However, do not adjust the hold control at this time. If you can't get the picture to lock in, check the capacitor C. It may be bad.

Now, take the short off the afc grid and the picture should lock right in. If it has a little trouble trying to lock in, adjusting the hold control should take care of the problem. If you still can't get it to lock in, it means that you have trouble in the afc section. Typically, there are two diodes in this section and it's possible that one of them may be bad.

Sometimes you can get the picture to hold but will lose sync. every time you readjust the horizontal hold. In this case, you're probably not getting horizontal sync. to the afc circuit from the sync. separator. Everything we have discussed so far can be done merely by watching the TV screen. However, if you have problems that originate in the sync. separator or afc, it's best to get out the scope. When using the scope, be sure that you have good, clean wave forms at each test point. For example, you should see a near perfect sawtooth at the afc diode. If not, check the sync. separator because some video is getting through with the horizontal sync. pulse.

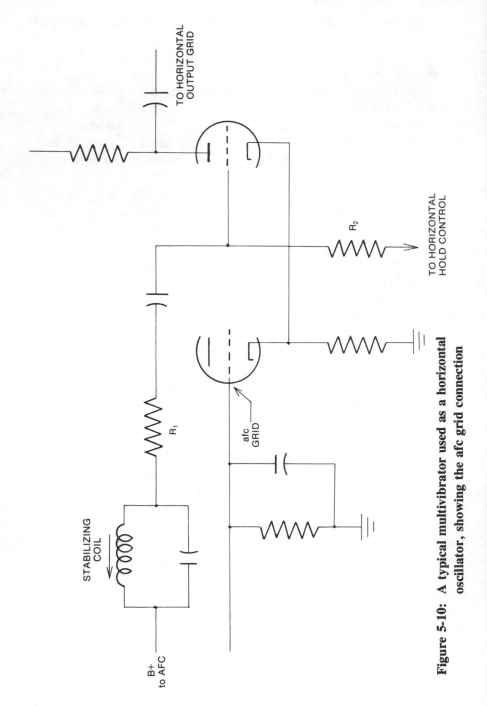

Figure 5-10: A typical multivibrator used as a horizontal oscillator, showing the afc grid connection

What You Should Know About a Horizontal
Output Amplifier's Adjustment

Anytime that a TV receiver needs a flyback transformer or horizontal output amplifier tube replacement, it's a good idea to check the cathode current of the horizontal output tube. In many cases, you'll find a removable link that enables you to place a milliammeter in the circuit without unsoldering.

If there is too much current, you'll have premature component failure. The higher the current is above prescribed values, the sooner the components are likely to fail. Generally, you should measure a cathode current of somewhere between 150 to 220 milliamps, depending on the type of receiver circuitry.

Probably the most common cause of high cathode current in color receivers is that the flyback circuit is detuned. In this case, see the service notes for the correct value, then while watching the milliameter (use the 0-500 mA scale) in the cathode circuit, tune the efficiency coil until you have the recommended value. If you can't lower the current to its recommended value by the adjustment, it's possible that you may find a shorted capacitor, screen grid capacitor, screen grid resistor, or a leaky component in the flyback circuit.

Practical Troubleshooting of a TV's
High Voltage System

While we are discussing flyback circuits, it's a good time to bring up the high voltage test. When the high voltage is too low, you will find the display can be too dark, hard to adjust, and possibly distorted. One way to make the test is to turn the brightness all the way up (this places full load on the high voltage power supply) and make and record a high voltage reading. Then turn the brightness all the way down (minimum load) and again make and record the reading. Now, there shouldn't be a large difference between the full load to minimum load condition. If the difference is great, it's an indication that there is trouble in the flyback system.

Any voltmeter is okay to make the test, provided that you use a good high voltage probe. However, be sure and remember that modern large color TV's have high voltages up to 30 kV. Therefore, your high voltage probe should be rated above 30 kV for these sets. If you simply want to check to see if there is a high voltage output, a good trick is to place a neon lamp on a long stick. Holding the stick, place the neon lamp near the cap of the horizontal output tube. If the lamp does not light, it means the high voltage system is not working.

How to Make a Successful Color TV
Sync. Section Alignment

The best way to align the sync. section (sometimes called the **AFPC**) is first to connect a color bar generator to the receiver antenna terminals and set the tuner to one of the low VHF channels. There are two things you should watch on the TV screen: One, the color of the bars indicates the phase of the 3.58 MHz oscillator and the stability of the oscillator is indicated by the stability of the bars. For example, if the oscillator is off frequency as little as 2 Hz, the bars will drift across the screen.

Your first step is to tune the color picture for best reception. Next, turn the color intensity control for the greatest color intensity you can get. Also, open up the color killer all the way. This will give you the greatest color intensity. Now set the color tint control to the center of its range.

After you have the setup, attach a high impedance voltmeter to one of the diodes on the transformer side in the phase detector. You should read a voltage somewhere between 20 and 40 volts and the polarity isn't important. However, whatever you read, write it down because you're going to need it. See Figure 5-11 for connections.

Now, locate the burst amplifier and place a jumper wire from its input to ground. Your next step is to adjust the 3.58 MHz oscillator output transformer for a maximum reading on your voltmeter. If you can get a good voltage, all is well and your oscillator is operating properly and correctly aligned. However, if your reading is abnormal you have a trouble in the oscillator or possibly in the detector coupling circuit.

Assuming everything is all right at the oscillator, take the short off the burst amplifier. When you do this, you should see the voltmeter reading make a sudden rise. What is happening is that the burst is now going to the phase detector and combining with the 3.58 MHz oscillator

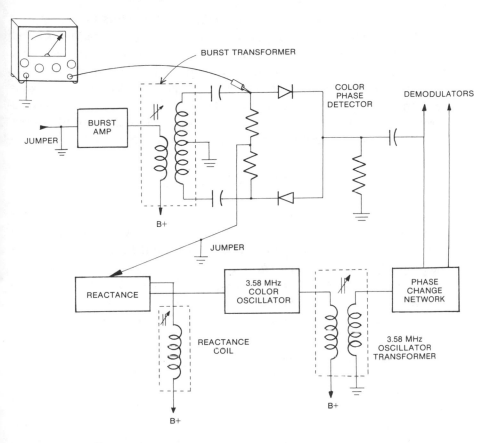

Figure 5-11:   Color sync. section alignment jumper and voltmeter connections

voltage. If you get a rise in voltage, it means that the burst amplifier is operating properly. If not, you'll have to troubleshoot it. Providing there is no trouble, adjust the burst transformer for maximum voltage. When you have maximum voltage, the transformer is correctly aligned.

If the receiver has a reactance circuit, you usually will find a test point where it is possible to place a jumper lead to ground it. When the reactance circuit is grounded, it places the 3.58 MHz oscillator in the free-running configuration. If your colors rapidly run across the screen, it is an indication that the oscillator transformer is quite a bit out of adjustment.

Adjust the reactance output circuit until the colors lock in or very slowly drift across the screen. Now, remove your grounding jumper and the colors should lock in. If not, check the 3.58 MHz oscillator and reactance circuit. Also, if you see the colors begin a fast whirl, check the diodes in the phase detector. Incidentally, all receivers do not have a reactance circuit. However, in case the receiver does have a reactance circuit and all you want is just a touch up, follow the procedure just described and you can have it done in just a minute or two.

Quick and Easy IF Trap Adjustments

Although we won't try to explain the alignment procedures for stagger tuned color TV receivers because it's better to have the manufacturer's service and alignment notes to do the job, we can explain the trap adjustments, which you can do quickly and easily. For instance, if you are experiencing herring bone interference on the TV screen, it is probably adjacent channel sound causing the trouble. To eliminate this problem, all you have to do is to locate a trap that is usually found in the first IF amplifier. Adjust this trap (while watching the TV screen) until the interference is at a minimum.

You'll also find another trap in the first IF amplifier, which is the adjacent channel video trap. Incidentally, the adjacent channel interference from the video is coming from the channel above, whereas the sound interference is coming from the channel below. At any rate, what you'll see on the TV screen is a windshield wiper interference and you want to try to tune this out by adjusting this trap.

In the output of the last IF amplifier, you normally will find a trap in series with a sound reject potentiometer. A problem that can occur here is that after the sound has been taken off, the carrier must be suppressed or it will beat with the color subcarrier and produce a 920 kHz signal. By watching the picture on the TV screen and adjusting the trap and sound reject, you can eliminate the beat. It's a good chance that you will see the color intensity improve at the same time because you're changing the response curve during the adjustment.

It is not suggested that you eyeball any other IF adjustments unless you are very familiar with the receiver circuits and have the correct equipment. In most cases, if you have never worked with the particular receiver, the manufacturer's service notes are a must.

What You Should Know About the Vertical
Interval Reference (VIR) Signal

A vertical interval reference signal for color is approved by the FCC for insertion in line 19 of the blanking period between television frames. The signal contains references for luminance amplitude, black level amplitude, and sync. amplitude, as well as chrominance amplitude and color burst amplitude and phase. At any point of origination—from videotape recorder to transmitter—the color, brightness, and contrast of the picture may be maintained at the proper level by restoring the reference signal to its exact specifications.

All three commercial TV networks, as well as the Public Broadcasting System, are inserting this reference signal in their transmissions, as are many videotape production houses and independent stations. Also, you'll find some TV receivers using this reference signal to automatically adjust chroma and tint. The sets have special circuits that automatically make the adjustments based on the station's transmitted VIR signal. A pilot light indicates when the VIR signal is present. However, the automatic correction may be overridden by a defeat switch (which is located on the front of the set) if you want to eliminate it during troubleshooting.

How to Effectively Suppress
RF Interference

This chapter is full of troubleshooting techniques that will be very helpful in solving one of the most common complaints heard by all radio and TV servicemen. How many times have you heard someone say, "I keep hearing a radio station on my hi-fi," or, "I keep hearing an amateur radio station on my TV."? Questions like these "bug" most technicians because the problem of suppressing interference can be a tough one.

Sometimes identifying the source can be half the battle for successful elimination. On the other hand, after you find the source you may not be able to do anything about it. In this case, there's nothing left to do but modify the receiver. However, there are certain techniques that can make the job easier. You'll find many practical suggestions and ideas on how to suppress interference, both at the transmitting and the receiving end, discussed in the following pages.

Tips for Identifying External RF Interference

One of the most frequent and troublesome types of interference that the technician encounters is external to the receiver. This is generally picked up in three ways: 1) by way of the antenna, 2) some part of the internal system acting as an antenna and another component acting as a rectifier (this frequently happens in tape decks, etc.) or, 3) through the commercial power lines.

Many times the interference is from rf transmitting equipment such as AM, FM, TV, amateur radio, and citizens band. In these cases, you frequently can identify the station by listening to the station call letters. However, identification isn't always that easy. For example, a few other potential noise sources and their characteristics are listed in Table 6-1.

NOISE SOURCE	CHARACTERISTICS	REMARKS
POWER LINE	LOW PITCHED FRYING SOUND	THIS IS CAUSED BY LOOSE CONNECTIONS, FAULTY INSULATORS, TREE LIMBS, ETC., DO NOT TRY TO MAKE CORRECTIONS. CALL YOUR POWER COMPANY.
NEON SIGNS FLUORESCENT ELECTRIC CLOCKS	BUZZING	FLUORESCENT LIGHTS INTERFERE AT 3.4 TO 8.3 MHz.
ELECTRIC MIXERS VACUUM CLEANERS MOTORS (SAWS) ELECTRIC SHAVER	STREAKS ON TV SCREEN	ANY ELECTRIC MACHINERY IS A POTENTIAL SOURCE. BY INSTALLING A FILTER CAPACITOR AT THE MOTOR, THIS CAN USUALLY BE REMEDIED.

Table 6-1: Potential interference generators

How to Pinpoint an Interference Generator

A practical approach to pinpointing an interference generator is to use a good field strength meter and a set of rabbit ears. Incidentally, if you don't have a field strength meter, try a small portable radio or TV receiver. They generally work quite well.

Before you jump off the deep end, it's a good idea to check and see if the interfering signal is coming from somewhere in your own shop or building. To do this, take your receiver or field strength meter to the main fuse box or breaker panel. Next, cut off the power to each circuit one at a time. If you're lucky, you will hear the interference stop when you pull one of the fuses or open a circuit breaker, and all that's left to do is check each piece of equipment on that line.

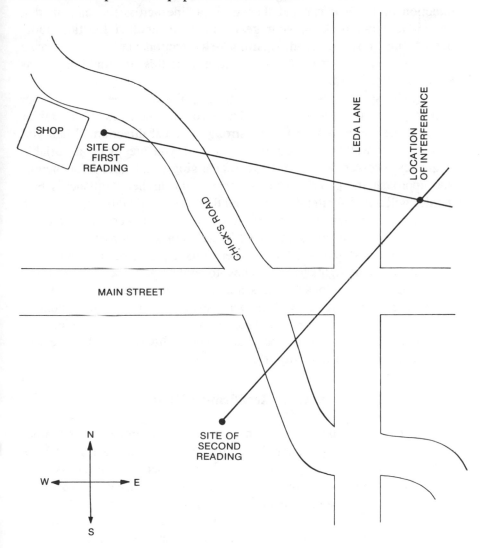

Figure 6-1: Using a map to locate an interference source

If your luck is like mine, the next step is to take a map of the area and mount it on a piece of light plywood. Now, with a compass, orient the map to line up with your area's actual north-south direction. Then during the time you hear the interference, take the rabbit ears and rotate them until you are receiving a maximum signal. Note the broadside direction of the antenna and then draw a line across the map in that direction. Now, pack up your gear and go to another location about one-half mile down the road. Again, take a directional reading and draw a second line on your map. After you have done this, it should appear as shown in Figure 6-1.

The system just described will pinpoint the general area. Your next step is to go to that location. Once you are there, rotate the rabbit ears, all the while looking for the strongest signal direction. When you find it, start walking toward the source. If you are using a variable frequency receiver, you want to keep changing to a higher frequency as you approach the culprit. This is because the higher frequency's harmonics will travel shorter distances and therefore, will help you zero in on the problem. For the same reason, it's best to use lower frequencies (if possible) when you first start your triangulation and on-foot search.

The last step is to ask the owner of the equipment to eliminate the interference. If he fights back, it's best to contact your local office of the Federal Communications Commission. However, remember that if you are in an industrial area very close to a transmitter or in a business zone of the city, the FCC permits a certain radiation level for all types of electrical equipment. Therefore, it may be your responsibility to filter the receiver, etc., or just learn to live with it.

How Audio Rectification Happens

One of the most troublesome types of interference is sometimes called *audio rectification*. Audio rectification has been defined by the EIA (Electronics Industries Association) as rf energy from any type of transmitter being received by an audio amplifier. Usually what happens is that one part of the audio system works as an antenna and another part —such as a transistor or integrated circuit—acts as a rectifier. Of course, once the rf signal is rectified, the following amplifiers will amplify the signal and it will appear at the speaker loud and clear.

If you are in a residential area, one of the most frequent sources of interference of this type is the amateur radio station. One of the simplest and quickest ways to locate the station is to look around the neighborhood for the station antenna. If you can't see the antenna (some amateurs place them in the attic), use your rabbit ears and a field strength meter as explained in the beginning of this chapter. However, after you find the station, it is more than probable that it will be operating within FCC specs and it then becomes your job to modify the audio equipment.

Troubleshooting Interference in Audio Systems

When you are checking the audio equipment, a good first test is to adjust the volume control for maximum attenuation. If you still hear the interference, it's being picked up after the volume control circuit. If you don't, it's coming in before the volume control. Now, if you find that it is being picked up ahead of the volume control, the next thing to do is to start a step-by-step check of all ground connections, solder joints, cables, and capacitors (especially electrolytics). After an electrolytic gets a few years old, it has a bad habit of developing a high internal resistance. Most servicemen have found that the easiest way to check these capacitors is to parallel a new one across the suspected one. While you are checking, watch for bad solder joints because they also can act as detectors. There-fore, anytime you find one that looks questionable, it should be re-soldered.

During your examination, look for any unshielded cables between chassis or being used for external speakers. If you find one or more in use, you can bet that they are acting as antennae. For this reason, it's a good idea to replace them with shielded cables. If the interference is after the volume control, you will have to place some type of rf filter in the following amplifier. It has been found that placing inductors in series with the base and/or collector leads of a transistor audio amplifier works best. However, it is important that you choose values that will not produce a significant change in gain or frequency response of the amplifier. The best way to check is to watch what happens to the signal on your scope when you insert the filter in the audio circuit. Ideally, there should be little or no change with or without the filter in the system.

Because you will have to use different values for your filter components on different products, it may be necessary to refer to a good reference handbook for filter design information. A handbook that I find very good is *Buchsbaum's Complete Handbook of Practical Electronics Reference Data,* published by Prentice-Hall, Inc., Englewood Cliffs, N.J. However, there are a few fairly simple methods explained in the next section you can try that, in many cases, will clear up the problem.

How to Deal with External Interference in Tube Amplifiers

An easy first step in dealing with a case of audio rectification is to try placing two disc type 0.01 microfarad capacitors from the speaker terminals to chassis ground, as shown in Figure 6-2.

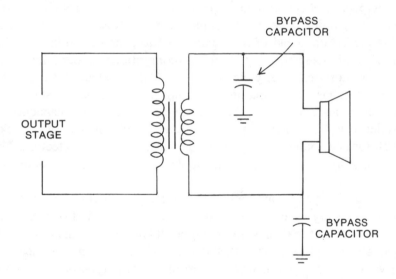

Figure 6-2: Simple speaker bypassing

Generally speaking, you are in pretty safe water when using capacitors to bypass a speaker. But, there is a "way-out" chance of capacitor loading, so if you have any doubts at all, it won't hurt to check with your scope. All you have to do is set the volume control at several different levels and look for any high frequency blips that might show up on the audio waveforms while music is being played. What you want to look for is a very high frequency audio oscillation that could possibly damage the tweeters in a hi-fi system. If adding the two capacitors doesn't work, you can go a step further. It may help to place a simple low-pass filter at the input to the first amplifier by adding a 75 k ohm resistor (R_1), and a 500 pF ceramic capacitor (C_1), as shown in Figure 6-3.

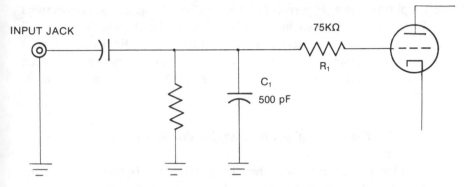

Figure 6-3: Adding a low pass filter to a tube type audio amplifier

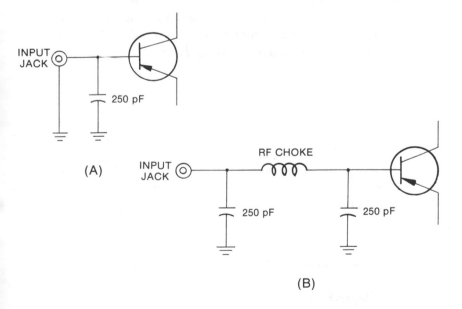

Figure 6-4: RF suppression in transistor circuits

How to Suppress External Interference in Transistor Amplifiers

A practical approach to suppressing rf interference in transistor audio amplifiers is to place a 250 pF ceramic capacitor between the input lead and chassis ground as shown in Figure 6-4 (A). Another successful technique that is a little more trouble (but usually more effective), is to insert a pi type filter. Figure 6-4 (B) shows how to make the connections and the values of the capacitors you will need. The rf choke's value will vary depending on the frequency of the rf interference. For example, try a 6 microhenry choke if the interference frequency is anywhere between about 30 to 90 MHz and about 1.5 micro henry for all frequencies from 90 up to about 200 MHz.

An Easy-to-Make Commercial Power Line Filter

It isn't unusual to find that the interference is coming in on the commercial lines. In some cases, simply placing a 0.1 microfarad capacitor across the lines may do the trick. If a single capacitor doesn't work, you'll have to get a little more serious and construct a more elaborate low pass filter. One that has been suggested by the EIA is shown in Figure 6-5. *Warning: All capacitors should be rated no less than 400 VDC and certified to be used across power lines.*

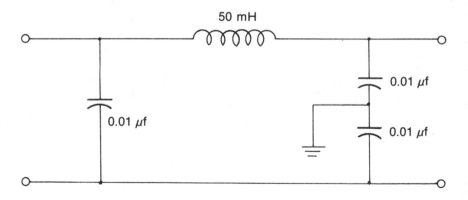

Figure 6-5: Commercial power line filter

Practical and Successful Mobile
Radio Interference Troubleshooting

If you ask a typical service technician to list the interference problems that have given him the most difficulty, there is a good chance that interference to a vehicle receiver will be somewhere near the top. It's easy to determine that it exists (just turn the receiver on with the engine running), but the problem is to find the cause and remedy. Many primary causes and possible remedies are listed in the following section. Incidentally, they aren't given in any particular order.

Cause 1. Breaker points and spark plugs
Remedy. Clean spark plugs and set correct gap for all spark plugs and breaker points. Also, check to see if all spark plugs have built in suppressor resistors. Your auto supply house can tell you if they do or don't. If they don't, replace them.

Cause 2. Generator brush sparking
Remedy. Check for an open by-pass capacitor across the generator. A fast way to do this is to shunt another capacitor that you know to be good across the old one. If the interference stops, it's the old capacitor that is in need of replacement.

Cause 3. Build-up of static electricity in tires and tubes while vehicle is moving
Remedy. This type of interference can be eliminated by connecting the wheels to the vehicle chassis. Even if you dislike the customer, it's not done by welding the wheels to the axle. It's done by placing small spiral springs inside the dust cap on the axle. Incidentally, the rear wheels are not a problem because they are fastened to the rear axle during manufacture. Also, it may help to use conducting powder inside the tires.

Cause 4. Long connecting wires and switches
Remedy. Bypass all long lines and switches. Switches generally will cause a popping sound in the receiver when they are activated. Therefore, they are easy to locate.

Cause 5. Antenna and transmission line
Remedy. Be sure that the antenna transmission line shield is at

ground potential. It is better to keep the antenna well away from the ignition system. If it isn't, it will probably help to move it.

A word of caution about work on ignition systems: A high voltage, which is in the vicinity of 10 kV, will be found at the coil, distributor cap, rotor, and spark plugs. Therefore, be very careful if you measure the voltage at these points because it's easy to lose a voltmeter. It's best to use an ignition analyzer if you wish to see and measure ignition waveforms. Or, use the old trick of drawing a spark by loosening a spark plug lead and holding it near the top of the plug. You should see a bright blue spark and the longer the spark, the higher the voltage. In other words, the same as you do on a "flyback" transformer in a TV high voltage system. If the spark appears weak and yellow, the problem probably is your condenser or coil.

Successful Shielding and Grounding

Shielding and grounding can make or break an electronics project. For example, we all know that placing an amplifier or instrument, etc., in a metal case will prevent external electrostatic charges from causing internal interference—which sounds very simple. And it is, provided it is done right. It would be nice if we could simply build a circuit, slide it into a metal case, slap a ground on the case, and say, "The job's done." But the real truth is that it just is not that easy. For instance, look at Figure 6-6.

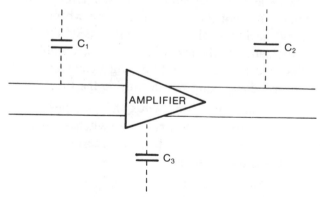

Figure 6-6: Electrostatic coupling inside a metal case

The capacitors C_1, C_2, and C_3, represent distributed capacities between the metal case and inside electronic components, such as unshielded cables, etc. What happens is that coupling takes place through the distributed capacities and the performance of the amplifier may be affected by external elements. All you have to do to get around this is short C_3 and you're in.

How do you do this? Easy. Just connect one signal line to the metal case. This generally is one side of the input or output line. In the case of the coax cable, the outside shield is normally grounded. Generally speaking, placing an earth ground on the outside case is solely for safety purposes and doesn't necessarily have anything to do with the signal path. However, you normally should select the signal lead that you want to ground so that it can be at the same point as earth ground.

In the case of the shield ground, it only needs to be connected at one end to establish ground potential. Two connections may add a ground loop and could very well cause a voltage difference to develop along the shield or between different chassis. This condition possibly could modulate the signal or, even worse, make you withdraw your hand rapidly when you touch it, leaving considerable skin behind.

Practical Techniques for
Improving the Antenna System

At one time or another, we all have run into the problem of having to match a 75 ohm line to a 300 ohm folded dipole or when interfacing test equipment, we found that their impedances were far from a match. One of the easiest and least expensive ways to whip this problem is to use a quarter-wave section of transmission line as a matching device. However, this is only one of the practical techniques you will find in this chapter.

Also included are practical, step-by-step procedures that will aid in improving the effectiveness of your antenna. For example, a foolproof way to get *maximum* power out of the antenna, how to measure antenna radiation resistance, and a simple but effective way to measure antenna gain, are all explained in detail.

Successful Transmission Line Matching

Frequently the technician must tune the antenna lead-in to a TV or some other high frequency device. With the TV receiver as an example, a neat and simple trick is to place a piece of TV lead-in as shown in Figure 7-1.

Here's how it's done. First, let's use channel seven's video carrier (175.25 MHz) where one-quarter wavelength is about 1.4 feet. In this case, you should start with a piece of lead-in of approximately 17 inches.

After you have connected the stub as shown in Figure 7-1, start trimming it little by little until you get a sharp, clear picture. Usually, it is better to short the cut ends each time and then twist the wire ends together

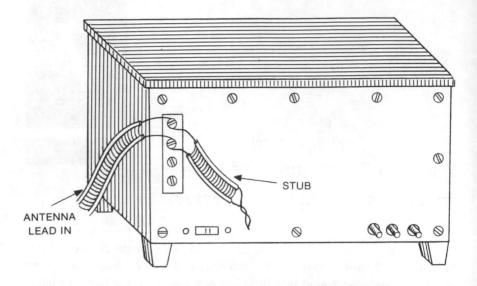

Figure 7-1: Using a stub to tune for maximum signal input

when you are finished. A word of caution: Placing a stub like this on a TV receiver may affect the reception on other channels. Therefore, you should check each channel for proper reception prior to wrapping up the job. However, normally this trick will serve you well when working with TV's, test setups, microwave shop work, or tuning antennas.

How to Construct and Use a Quarter-Wave Matching Transformer

Another good approach is to use a section of transmission line as a quarter-wave matching transformer. The "hook-up" is shown in Figure 7-2.

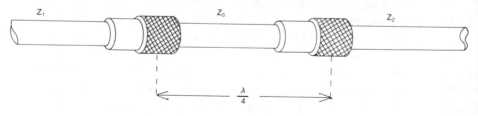

Figure 7-2: Using a quarter-wave matching transformer

Start by placing the matching section (Z_0) between the two mismatched transmission lines (Z_1, Z_2). Incidentally, before we go too far you should realize that, as before, the load (Z_2) can be an antenna, a piece of test equipment, or a receiver. It is also important to know that you can't get away with the cut-and-try method here because it is necessary to know the value of the characteristic impedance of the matching section.

A very simple calculation will reveal this secret. For example, take the problem of matching a *nonreactive* 73 ohm line to a 125 ohm line. This will produce an SWR of about 1.7 to 1. What you need is a quarter-wave piece of 95 ohm line to reduce this SWR to approximately one-to-one. How do we know this? Easy. Just use the formula,

$$Z_0 = \sqrt{(Z_1)\,(Z_2)}$$

where the characteristic impedance (Z_0) is the impedance of the line you want to use for a matching transformer and Z_1, Z_2 is the characteristic impedance of the two lines you wish to match. All that is left to do is insert the values into the formula.

$$Z_0 = \sqrt{(73)\,(125)} = 95.5 \text{ ohms}$$

You might use a section of RG-U-57A (twin conductor), which has the desired impedance (95 ohms). However, the type of line used would depend on whether you need low loss, low capacity, miniature coax, etc.

When working with a receiver and its antenna system, you easily can tell when you have a good impedance match because you will have the best reception. However, the transmitter and its antenna are not as easy. Here's how to whip this problem.

Key Steps to Loading the Antenna

Remember when your instructor made the seemingly simple statement, "For maximum power transfer, the source must be matched to the load ohm for ohm."? Well, it didn't take long for all of us to find out that it was easy for him to say and difficult for us to do. We quickly discovered that what we had to know was whether the antenna appeared capacitive, inductive, or as a pure resistance. The resistor substitution technique will tell you all this and would have satisfied your instructor,

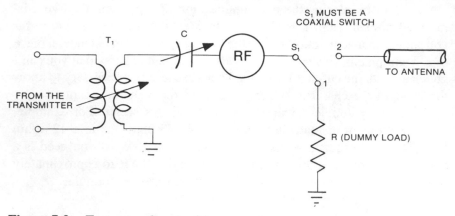

Figure 7-3: Test setup for checking an antenna

too. Figure 7-3 shows the test setup and the following procedure tells you step-by-step what to do.

Here's the procedure:

Step 1: Begin by placing the coaxial switch (S_1) in the dummy load position (1). Next, adjust the coupling of T_1 until you read approximately mid-scale on the RF ammeter.

Step 2: Now, tune the capacitor for maximum reading on the RF ammeter.

Step 3: Place switch S_1 to the antenna position (2). While watching the RF ammeter, tune the capacitor until the meter reads maximum current and take note of whether you had to increase or decrease the capacitance.

Once you know this, you're in. If you had to increase the capacitance, the load is capacitive. If you decreased it, the load is inductive. Or better yet, if there was no change required, the load appears as a pure resistance. However, most of us are not that lucky on the first try. Therefore, you probably will have to tune the antenna, adding more inductance or capacitance in series with the feed line. If it appeared capacitive, you need more inductance. If it was inductive, you need more capacitance.

When you have the antenna properly tuned, switching between the antenna transmission line and the dummy load will make no change on the RF meter. In other words, the antenna is properly tuned for maximum power output.

It's also quite possible to tune for maximum smoke if you don't select the proper capacitor plate separation and the right RF ammeter. This will depend on the maximum voltages and currents you expect during the measurement. Furthermore, the dummy load should be non-reactive and capable of dissipating the applied power. However, I have used common light bulbs for a dummy load when tuning up medium power transmitters. Two or three one-hundred watters work nicely. The values of the capacitor and inductor can be calculated by using the following formulas,

$$C = \frac{0.025}{(fr)^2\,L} \quad \text{or} \quad L = \frac{0.025}{(fr)^2\,C}$$

where **fr** equals the operating frequency you are interested in.

For example, if you happen to have a coil of 60 microhenrys on hand and want to tune an antenna working at a frequency of 649 kHz (in the broadcast band), the capacitor you would use is about 1,000 pico farads.

There is another value that is nice to know (and sometimes you must know) when working with the antenna. This is the antenna radiation resistance. Why do we want to know this? Well, for one reason, with this value it's easy to calculate the *true* power output of your antenna. To find this elusive value, here's what to do.

A Practical Way to Measure Antenna Radiation Resistance

To keep things simple, this technique will use the same components that were used in checking antenna loading in the last section with only a variable noninductive resistor being added. The circuit connections are shown in Figure 7-4.

First things first. Start by assembling the equipment, which will include the following:

1. An rf signal source—the power output of the signal source should be large enough to produce at least a mid-scale reading on the milliammeter described in item 3.

2. A variable and noninductive resistor

3. A radio frequency milliameter that reads from 0-1000—the

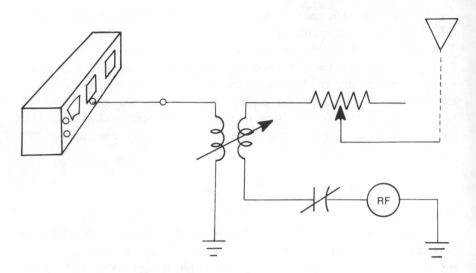

**Figure 7-4:   Circuit diagram for antenna radiation re-
sistance measurement**

full scale reading of the meter will depend on the maximum
amount of power that is to be used to excite the antenna.

4. A variable inductance of 60 microhenrys.

5. A tuning capacitor of approximately 1,000 pico farads.

Before you start the actual measurement, it is best to let the signal
source warm up for about an hour or so. After you have done this, set the
source to the desired frequency and loosely couple it to the antenna. Now,
to continue, make each of the following adjustments.

1. Adjust the capacitor until you find resonance. This will be
indicated by a maximum reading on the RF ammeter.

2. Readjust the coil coupling until the RF ammeter reads half-
scale (500 milliamperes, in this case) with the resistor set at
zero ohms.

3. Now, increase the resistance. When the RF meter reads exactly one-half of its previous value—where you read 250
milliamperes—the value of the resistor is equal to the antenna
radiation resistance at applied frequency.

You will notice that this is the radiation resistance at the carrier frequency, which raises the question, "How about the case of a modulated carrier?" The answer to this question is that when you are dealing with sidebands, you should take readings at several points on each side of the carrier frequency. For example, in the case of an AM broadcast station this is done at 5, 10, 15, and 20 kHz on each side of the assigned carrier frequency.

These measurements are then plotted on rectangular graph paper with the frequency placed on the X-axis and the resistance reading on the Y-axis. A curve is drawn so that it follows the plotted points and where the curve intersects the operating frequency is the correct value of antenna resistance.

How to Determine Antenna Power Output with an RF Ammeter

Once you have determined the antenna radiation resistance as just explained, you're ready for the finishing touch. Place an RF ammeter in the line to the feed point of the antenna and use $P = I^2R$ to calculate the antenna power out. Radiation resistance is R, antenna current is I, and radiated power is P. Incidentally, once you have made this radiation resistance measurement, it does not change unless there is a physical change in the antenna or in the operating frequency. Now, take a reading of the antenna current and just punch in I^2R on the old pocket calculator and you have the output power in a split second. And it's this easy regardless of how many times you change the power at the transmitter.

How to Measure Antenna Gain

One of the problems you run into after constructing a directional antenna is measuring the gain. Here is a fairly simple way to do the job that should make life easier. A practical measurement system that you can use is shown in Figure 7-5.

Begin by placing the receiving antenna in the far-field region (at least five wavelengths distance from the transmitting antenna). Next, the receiving antenna should be positioned to obtain maximum received power, as indicated on your indicator. The indicator can be a vacuum tube voltmeter. However, if you are working with microwave equipment, a standing wave meter is better. Now, set the variable attenuator to zero dB

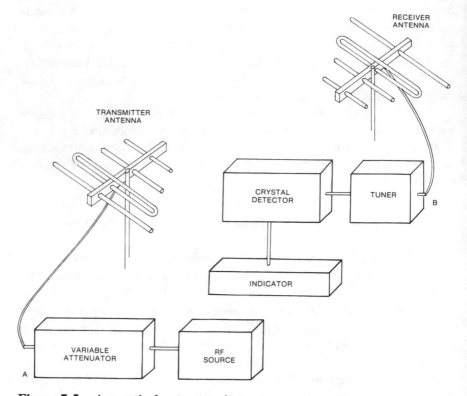

Figure 7-5:   A practical antenna gain measurement system

and adjust the indicator to about mid-scale by use of the appropriate controls. Once these adjustments are made, *do not* change the settings of anything except the variable attenuator during the following procedures.

Remove the tuner, crystal detector, and indicator (a suitable radio receiver and VTVM can be used here) from the receiving antenna at point B and carry it to the transmitter site. The next step is to set the variable attenuator at the transmitter to a high enough level to prevent any damage to the tuner. (The output must be below the original reference level.) After you do this, disconnect the transmitting antenna and connect the measuring system to point A.

Now, adjust the attenuator until you have the original reference level on your indicating instrument. The attenuator reading in dBs is the familiar power ratio $10 \log P_2 / P_1$. To show how to use this reading, let's assume that you know the gain of the transmitting antenna. The gain of the receiving antenna can be found by using the following formula:

G_r = 22 + (20 log **R**) − (20 log λ) − the difference in the attenuator reading in dBs − gain of the transmitting antenna

> Where: G_r = receiving antenna gain in dBs
> R = separation of antenna in feet
> λ = free space wavelength in feet

An Example of How to Calculate Antenna Gain

To show how to calculate antenna gain, let's assume that you have two antennas spaced 60 feet apart. The frequency you are working with is 9 GHz. Calculating the wavelength results in 0.1092 feet. Next, your attenuator difference reading is 28 dB and the transmitter antenna gain is 24.4 dB. Now, if your calculator does logs, it's simply a matter of pressing keys. Your work would look like this:

> **G** = 22 + (20 log 60) − (20 log 0.1092) − 28 − 24.4
> = 22 + (35.56) − (−19.23) − 28 − 24.4
> = an antenna gain of 24.39 dBs

Another point: Don't forget the rule, "When subtracting a minus number from a positive number, always change the sign of the subtrahend and proceed as in addition." In the above problem, 22 + 35.56 − (− 19.23) equals 76.79. In other words, just add them up.

One final thing: Most of the time transmission line loss can be disregarded. However, if you are working with long lines (such as 100 feet) and the operating frequency is 100 MHz or above, it's better to include the line attenuation in your calculations.

Troubleshooting and
Servicing Recording Equipment

This chapter deals with techniques that will not only help you get recording and reproducing equipment out of the shop faster, but also in many cases with superior performance. Servicing this type equipment has undergone a phenomenal growth in the past few years and it will continue to increase, according to all statistics. In view of this rising trend, knowing the tricks and techniques used to repair this equipment could put you ahead of the competition.

Troubleshooting and Servicing Tape Recorders

When you troubleshoot a tape recorder, you make the preliminary checks of the electronics components just like any other electronics circuits. In other words, look for burned resistors, bulging capacitors, cold solder joints, broken or loose connections, excessive heating, make ohmmeter checks, voltage measurements compared to manufacturer's instructions, and the like. However, how to service the heads and mechanical parts is a mystery to many electronics technicians. Unless you keep on top of the recording field, it's very probable that you won't be familiar with what to look for when making preliminary checks on equipment of this type. Figure 8-1 shows a typical mechanical arrangement you might run into.

Before we go too far, it should be pointed out that the very first

COVER REMOVED

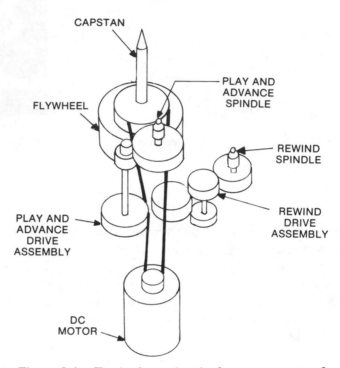

Figure 8-1: Typical mechanical components of a battery-operated cassette tape recorder

step should be a good, thorough cleaning of the transport system, especially if you are experiencing poor playback, hum, or poor erasure. Problems like these generally are due to oxide that has transferred from the tape to the heads, guides, pressure rollers, capstan, or pressure pads over a period of time.

Electronic supply centers (such as Radio Shack, etc.) sell cassette demagnetizers that also remove oxide residue and kits for cleaning and lubricating all types of recorders. But sometimes it isn't possible to get to one of these supply houses for one reason or another. If this is your problem, a good solution is to use a cotton swab and some rubbing alcohol. Both can be purchased at any drugstore. In order to remove oxide buildup using a cotton swab and alcohol, follow this procedure:

> *Step 1.* Dip your cotton swab into the alcohol, making sure that you don't oversaturate. It's best to squeeze the swab almost dry before doing the next step.
>
> *Step 2.* Swab each head, roller, etc. very thoroughly until the oxide residue is removed. You can use another dry cotton swab to remove any fluid that may be left.
>
> *Step 3.* An easy way to clean the capstan is to turn the machine on and hold a wet cotton swab against the rotating capstan. However, in the case of cassette recorders, you may have to actuate the machine internally to make the capstan turn.

A Handy Tape Recorder Visual Checklist

Always visually check a recording machine before you do any extensive disassembly. Here is a list of visual symptoms that indicate there is some trouble:

> 1. Look for off-center movement as the tape passes the capstan. If the tape is off center, you need to adjust the capstan and/or pinch roller. Because there are several methods used for the adjustment, it's best to consult the service manual for the machine you are servicing.
>
> 2. Check to see that the tape is making proper contact with the heads and that there is no oxide build-up or head misalignment.
>
> 3. Check to see if the head is worn unevenly or has cuts, nicks, or scratches. If any of these conditions exist, the head should be replaced.
>
> 4. Use a magnifying glass and check the tape gaps on the heads. (You can't see them with the naked eye.) You should see very fine lines, each the same width. The erase head lines will appear wider than the record/playback heads. But in any case, on any single head they should appear as all having the same width or the head should be replaced.

5. Check for proper lubrication. *Do not overlubricate.* Generally, once a year is sufficient. Only use silicone base lubricants on the heads or anywhere that grease might come in contact with rubberized surfaces.

Troubleshooting Recorder Heads

Every electronic communications technician knows that the higher the frequency, the shorter the wavelength. However, many don't know that the same thing holds true when dealing with magnetic tape. That is, the higher the frequency, the shorter the marks on the tape. The important point is that even the slightest misalignment of the heads will affect the high frequency response of the recorder, as well as cause interchannel cross talk on a multichannel machine. There are five basic head adjustments that you must be familiar with to eliminate these and other problems.

Vertical position (Tilt): Generally, the first step is to align a head for true vertical in reference to the tape. Check to see that the head doesn't lean into or out from the surface of the tape. See Figure 8-2, which shows a case of head vertical tilt.

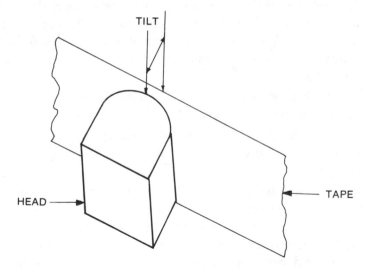

Figure 8-2: Tape recorder head with vertical tilt

One indication that you may see if the head is out of vertical alignment is that the tape will run a little high or low in reference to the center line of the tape path. Another possibility in the case of the erase head is that you might hear part of a previous recording in the background of a new recording.

> *Head Height:* Next, check to see if you have proper head height. See Figure 8-3, which is a case of improper head height adjustment.

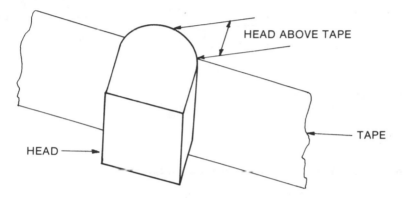

HEAD ABOVE TAPE

TAPE

HEAD

Figure 8-3: Improper head height adjustment

If you hear cross talk in a multitrack recording or you can't make a tape run on another machine, it may be because a head is a little high or low and needs adjustment. The cure is an obvious one, however, as always, it's best to refer to the service manual to make the adjustment. An example of one type of head height adjustment is shown in Figure 8-4.

> *Horizontal Alignment (Tangency):* What you want in this case is to be sure that all heads are perfectly square to the tape surface. See Figure 8-5, which shows a head with improper horizontal alignment.

The symptoms may be a loss of small bits of information, a lower signal amplitude, or you may hear another recording in the background if it is an erase head trouble. If it is a playback or record head problem, it's possible that you may lose the entire signal if the head is drastically out of line.

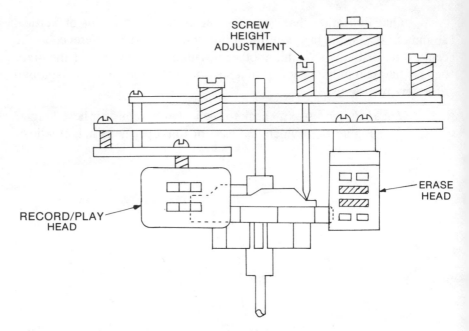

Figure 8-4: Mechanical head height adjustment

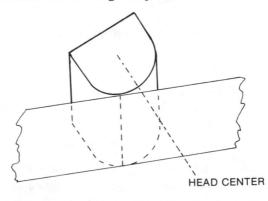

Figure 8-5: Improper head horizontal alignment

Azimuth Adjustment: This adjustment is probably the most important of all, especially if different heads are used for recording and playback. Figure 8-6 is an example of both correct and incorrect azimuth alignment.

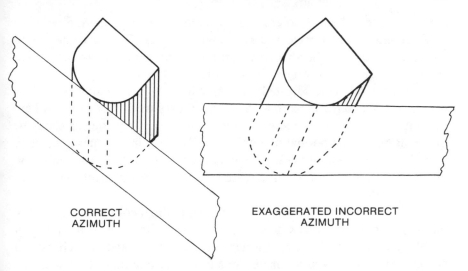

CORRECT
AZIMUTH

EXAGGERATED INCORRECT
AZIMUTH

Figure 8-6: Alignment of head azimuth

The objective during this check or adjustment is to be sure that the head gaps are exactly at right angles to the path of tape travel. In reality, you can get away with some azimuth error if the recorder uses one head for both record and playback. But, it's best to keep all heads in proper adjustment because you may encounter problems when you try to run the tapes on other machines.

Easy-to-Make Head Azimuth Check

To make a quick check to see if the azimuth is in need of adjustment, simply use a polished toothpick to move the tape *gently* up and down across the head while the machine is running in the playback mode. If this produces an increase in the output level, the azimuth is in need of adjustment. However, if you hear a decrease, it's an indication that the adjustment is okay as is.

How to Make an Azimuth Adjustment

The first thing that you should do before starting to actually make an azimuth adjustment on a normal three-head recorder or a two-head recorder (where there is no erase head) is to obtain a standard alignment tape recommended by the recorder manufacturer. The next step is to adjust the playback head azimuth for maximum output as read on a suitable meter (VU or dB) while the alignment tape is being played.

After you have adjusted the output level to maximum, take off the standard tape and then record a 15 kHz tone on a well-erased or new tape running at 15 inches per second. You should record the tone at about 10 or 20 dB below the normal maximum for the machine. Your next step is to observe the output of the head and adjust the azimuth adjustment screw on the *record head* until you have a maximum output level reading on your dB meter.

Now, with the machine still running in the recording mode, check the output on the playback head again; it should be okay. If it isn't, you probably didn't make the adjustment of the record head as well as you should have. After you've made the adjustment one more time and you still have trouble, replace the head (or heads) and redo the adjustment again, which should complete the job.

How to Check the Speed of a Turntable

When using the following system, there are two things that you will need to check the speed of a turntable. One is a stroboscope disk and the other is a neon or fluorescent light, which must receive its energy from the 60-cycle commercial power line. On the disk, you will find several separate circles with a different number of bars on each circle. For example, if you want to check the speed of a 33 1/3 rpm turntable, you would use the ring that has 216 bars. Or, for 78 rpm, you would use the 92 bar circle.

The procedure is to shine the light on the turning disk and if you see the bars appear stationary, the speed is correct. In general, you should not see more than 21 bars a minute drift past a selected reference point in either direction. You can determine whether the speed is too fast or too slow by watching which way the bars drift. If the bars drift in the opposite direction to the way the disk is turning, the speed is too slow. If the drift is in the same direction, the speed is too fast. Professional turntables usually will have an adjustment to correct the problem and all you have to do is turn the vernier control until the bars appear stationary. If there isn't an adjustment, it's best to check the drive mechanism for worn parts.

How to Check the Frequency Response of
Turntables and Tape Recorders

One great advantage of the following frequency check when made on a tape recorder is that it automatically checks tape head alignment, frequency response of the amplifiers, equalizing circuits, bias oscillators, and tape head performance. To check all of these on a tape recorder recording unit, proceed as follows:

First: Using a very good quality audio signal generator, record several frequencies between 50 and 20 kHz, maintaining a constant level input at each frequency you choose.

Second: Attach an external VU meter to the machine if it doesn't have one built in. Then play back the tape and record the meter reading at each frequency that you have recorded.

Third: Draw a graph using the readings you have written down. The vertical line should show amplitude, and the horizontal, frequency. The result will be the frequency response of the tape recorder recording system. Your plot should look something like the one shown in Figure 8-7.

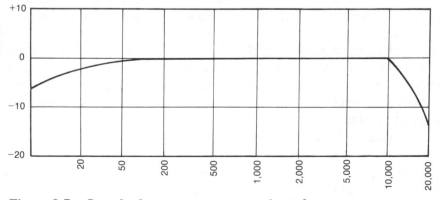

**Figure 8-7: Sample frequency response chart for a
tape recorder**

You can check the frequency response of a turntable pickup by using a standard EIA frequency test record. However, in all probability you will have to use an amplifier to raise the output signal of the pickup to a usable level. And, as we all know, any amplifier will have a frequency response of its own. Therefore, you first must measure the frequency response of your test amplifier. Use exactly the same procedure as was described for checking tape recorder frequency response. That is, place a VU meter on the output of the amplifier and plot a curve showing frequency vs amplitude.

Next, connect the amplifier to the turntable pickup output and make a new graph using the same test frequencies as you used to check the amplifier. But remember, in all cases you must maintain a constant signal level output from your audio oscillator during the recordings. Now, with the frequency response curve of the amplifier and the frequency response curve of the overall system, you can compare one against the other and determine the response of the turntable pickup alone.

How to Use a Square-Wave for Testing

If you just want to make a quick and easy test of a transistor audio amplifier, etc., all you need is a scope, high impedance probe, and a square wave generator. The first thing you should do is to check the value of the capacitor on the signal input lead of the amplifier you want to test. Next, to prevent distortion and for the safety of the amplifier under test, place a capacitor of larger value between the signal generator and the amplifier.

Before you test an amplifier for square wave response, it is advisable to check the appearance of the square wave at the highest and lowest frequency you intend to use. This is because the oscilloscope itself may degrade the waveform. If it does, all measurements must be made with this fact in mind.

After you've checked your scope, set your generator to the lowest frequency you want to use and connect your scope to the amplifier output. Raise the square wave generator's output in steps of 10 kHz at a time up to about 15 or 20 kHz, all the while looking for any excessive distortion. In all testing like this, you'll observe some distortion but if the square wave remains basically the same at all times, the amplifier can be considered to be linear. Figure 8-8 displays several possible ways a square wave could appear on your scope. Figure 8-8 (1) shows a perfect square wave that indicates that the amplifier is not distorting. Figure 8-8 (2) shows the leading edge rounded off. This indicates that there is a loss of gain at high frequencies. If you observe a substantial drop at a certain high frequency, the eighth harmonic for instance, and assuming a 2 kHz square wave is placed on the input without rounding off, the amplifier is linear to 8×2 or 16 kHz. Figure 8-8 (3) indicates that there is a higher gain at mid-frequencies. Figure 8-8 (4) is a case of differentiation. This can happen when the grid resistor or coupling capacitor is not the correct value. For example, the coupling capacitor is not large enough or it could be a partially open capacitor.

Figure 8-8 (5) shows a square wave with dampened oscillations. The cause of this trouble could be that distributed capacitances and inductance are at resonance at some low frequency. The cure for this usually is to shorten your leads, or sometimes just moving the leads around will help. If you are working with video equipment, it may be that a peaking coil needs a damping resistor across it.

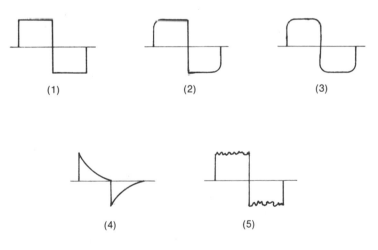

Figure 8-8: **Oscilloscope patterns of a distorted square wave that may be observed during testing of an amplifier**

How to Measure Phase Delay with a
Dual Trace Scope

One of the easiest ways to check for phase delay is with a dual trace scope. Your first step is to apply a sine wave to the input of the amplifier and monitor it on one channel. Next, monitor the output of the amplifier on the other channel. Now, you should see a time differential between the two waveforms if there is any phase delay. The measured time-delay, divided by the period of the input signal, multiplied by 360 degrees will give you the phase shift in degrees. If you can't get a uniform phase shift across the amplifier bandwidth it's an indication that the amplifier has an irregular frequency response.

How to Successfully Troubleshoot, Test, and Align, AM and FM Equipment

This chapter concentrates on successful ways to get results when dealing with the many different problems encountered when working with AM or FM equipment. It shows how to determine the power requirements of an AM modulated amplifier, measure percent of modulation, and calculate tube power dissipated, just to name a few of the things you'll find.

However, the technician also must be familiar with what to do about receiver troubles, such as an AM or FM broadband receiver being out of alignment or troubleshooting the all-important front end. You'll find out how to solve these troubles as well as others, all of which help you to become a more skilled and sought-after electronics technician.

How to Make a Broadband AM Alignment

Many technicians believe that an AM broadcast station is restricted to a 5 kHz modulating signal. This used to be true a few years ago—but not today. In fact, many AM broadcast stations use modulating audio signals up to 15 kHz. Furthermore, many modern receivers have a bandpass capable of passing the new channel width. Now, where the trouble comes is when some unknowing technician makes a standard band (10 kHz) alignment in these receivers. If you run into a trouble like this it's better, and much easier, to use a sweep generator and scope to align receivers of this type.

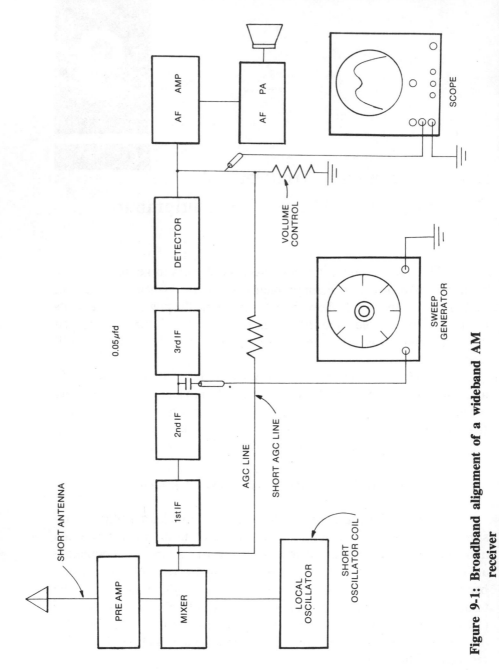

Figure 9-1: Broadband alignment of a wideband AM receiver

Most of these receivers will have at least two IF stages and some-times more, and what you want to see on your scope after completing alignment of all stages is as close to a perfect square wave as possible. However, it's a bit tricky to accomplish this. To start the job, the first thing you have to do is disable the receiver's local oscillator, AGC, and short the antenna. The reason you want to do this is to prevent the receiver's local oscillator from heterodyning with the sweep generator, stop all signals from being picked up by the antenna, and keep the au-tomatic gain control from attenuating the test signal.

The easiest way to disable the local oscillator is to place a jumper lead across the tank coil. Then to disable the AGC, look at the receiver schematic and find the AGC bias feedback line and short it to ground. After you have everything listed disabled, attach the sweep generator and scope as shown in Figure 9-1.

Your next step is to place a 0.05 μ fd capacitor on the sweep generator test lead and connect it to the input of the last IF amplifier. This point will be the grid if it's a tube type amplifier, or the base if it's a transistor amplifier. Then, place your scope at the output of the detector. Across the volume control is normally an easy place to make the connec-tion.

Now, turn the generator, scope, and radio on and let them warm up for at least a half-hour. When I have work to do with test equipment such as this, I either turn the equipment on at quitting time the day before and leave it on all night or turn it on the first thing on coming to work. If

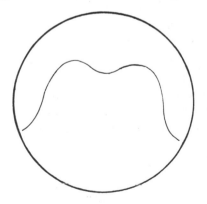

**Figure 9-2: Proper response curve of a wideband AM
receiver IF amplifier**

you turn it on early in the morning, I've found it a good idea to wait until after coffee break (about 10 o'clock in most shops) if possible, to do the job. After everything is warmed up, adjust the sweep generator for a sweep of about 15 kHz. You should see a trace on your scope like the one shown in Figure 9-2 if the IF transformers are properly aligned. However, if the IF transformers are misaligned, you can see a variety of patterns on your scope. Three possibilities are shown in Figure 9-3.

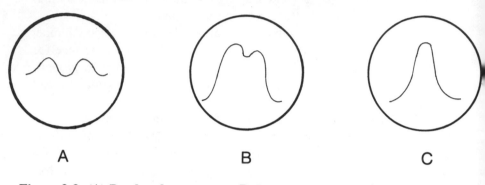

A   B   C

Figure 9-3: (A) Passband too narrow (B) Improper coupling, needs additional adjustment (C) A case of overcoupling

To make the alignment, adjust the secondary of the last IF amplifier, and then do the primary next. As you make each adjustment, try to keep the curve on the scope looking like the one shown in Figure 9-2. In order to produce and keep a curve like this, the IF amplifiers are frequently stagger tuned. For example, one single bandpass is only about 5 to 8 kHz wide. However, by stagger tuning, you can get up to about 15 to 16 kHz and the result is as shown in Figure 9-4.

Now, when you have a good-looking pattern on your scope, move the sweep generator back to the next IF transformer and adjust as necessary. Do the secondary first and the primary second. Keep doing this right on back to the first IF amplifier. One word of caution! Don't try to go back and realign an IF amplifier transformer without moving the sweep generator to that input because if you don't, you may have to start all over again. Incidentally, sometimes you'll have overloading due to the fact that the local oscillator is disabled. If you do, simply reduce the signal output of your sweep generator and this should alleviate the problem.

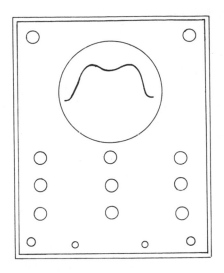

**Figure 9-4: Pattern on an oscilloscope due to stagger
tuning**

Receiver Front-End Check and Alignment

A quick way to find out if the local oscillator of a receiver is out of adjustment is to check the dial setting of the stations. Generally, as a first step it's best to check the readings on the dial against the actual frequencies of two stations at opposite ends of the dial. In most cases, you'll find that the local oscillator frequency is the sum of the received station frequency and receiver intermediate frequency. There are three simple formulas useful during front-end alignment, which are as follows:

$$LO = RF + IF \qquad RF = LO - IF \qquad IF = LO - RF$$

where LO is the local oscillator frequency, RF is the carrier frequency of the station you are tuned to, and IF is the intermediate frequency of the receiver.

To make a front-end alignment, first inject the signal generator signal into the antenna terminals. If the antenna is the loop type, simply place the generator output lead next to the antenna. This normally will produce ample signal input. However, if you make a direct connection to the antenna terminals, don't forget to use a standard blocking capacitor in the signal generator input lead. A 470 picofarad capacitor should do the trick. Also, a good place to connect your scope is across the volume control, as mentioned earlier.

Now, tune your AM receiver to the top of the AM band somewhere near 1400 kHz. Next, set your signal generator to the same frequency. What you want to do now is adjust the *local oscillator trimmer capacitor* (it is across the oscillator coil) for maximum output as seen on your scope. Your objective is to set the oscillator to exactly the correct frequency at the upper end of the dial.

The next step is to tune to the lower end of the dial (about 600 kHz should be a good spot) and repeat the procedure. In this case, you want to adjust the *RF trimmer capacitor*. You'll find it in parallel with the RF tuning capacitor. Also, set your signal generator at 600 kHz if that is the frequency you chose to use. Finally, after the two adjustments, you should find that the receiver will track (i.e., each station in the right place on the dial) from the low end to the high end without any trouble in between. Almost always, the front-end should be aligned anytime an IF alignment is done. But, be sure that all shorting leads are taken off the AGC, local oscillator, and the antenna if you have to make front-end adjustments.

How to Improve Ferrite Core Antenna Sensitivity

If you have a ferrite core antenna and just can't get the sensitivity you want, it might help to adjust the antenna. To do this, you have to make two adjustments. (1) Adjust the RF trimmer capacitor, which is usually across the RF tuning capacitor, and (2) move the antenna wire on the ferrite core. You will find that the antenna wire is cemented to the ferrite core. Therefore, you have to use a suitable solvent to free it. Once you have it free, tune your receiver to a station at the low end of the dial (about 600 kHz, if you are using the AM band) and move the core by hand until you have maximum signal output on the receiver.

Next, tune up to somewhere near 1400 kHz and adjust the trimmer capacitor for maximum output. Do this several times until you have a peak output at both ends of the dial and then, being careful not to move anything, recement the antenna to the core.

It should always be remembered that the front end of any receiver is the most critical of all with respect to receiver sensitivity. Therefore, the antenna, mixer, and local oscillator always are the first suspects when dealing with tracking and sensitivity troubles.

How to Use Noise in Testing

Many technicians are not aware that noise can be a handy tool for fine tuning the front end of a receiver. One simple noise source is the nearest fluorescent light. The noise is a hiss in the receiver and you'll find it, for all practical purposes, across the whole RF spectrum. To use a fluorescent light, set the volume control at maximum and then use the distance between receiver and light to adjust for the desired signal level. Although this isn't the best way in the world, you can use it if you're in a bind.

Your next step merely is to set the receiver at both the low and high end of the dial and adjust for maximum static. At the low end of the dial, adjust the oscillator trimmer and at the high end, adjust the antenna trimmer. If you would like a better and controllable noise source that will do a much better job, construct the simple circuit shown in Figure 9-5 and follow the procedure given in the next section.

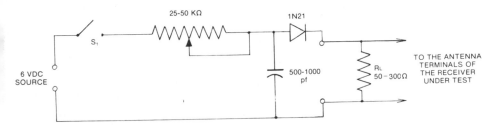

Figure 9-5: Simple adjustable noise generator

Copyright © Ziff-Davis Publishing Company
Reprinted by permission of Popular Electronics Magazine

The value of the load resistor (R_L) shown in Figure 9-5 should be the same as the input impedance of the receiver you're testing. Also, it's best to use a very well-regulated power supply or a battery for the 6 VDC source. If there is any doubt about your power supply, the battery is the better of the two.

The first step is to disable the AGC and then connect the noise source to the antenna terminals with switch S1 open. Next, connect an AF voltmeter that has a dB scale to the receiver output. Now, turn the receiver on and adjust the gain control (or gain controls, if there is an RF and audio gain control) until you have a convenient reading on your dB meter. This is your reference noise level. Place switch S1 in Figure 9-5 to the on position, which activates the noise generator, and adjust the 25 – 50 k ohm pot until you get a minimum discernable increase on the dB meter. This is roughly your receiver threshold, in other words, the weakest signal the receiver can detect.

Again, this isn't the most accurate test in the world due to the diode being used. However, it can be used to do a pretty good job of fine tuning a receiver front end. In this way, you can find which resistor, transistor, etc. is creating the greatest noise and replace it. What you want is to keep trying to increase the dB reading with the noise generator on and decrease the dB reading with the noise generator off, but don't change any of the gain controls during your work. Also, be sure that you're not tuned to a station or other signal source during the tests and adjustments because if you are, all readings will be in error.

An Improved FM Demodulator Alignment

A common trouble found in FM radios and other FM equipment is that the discriminator or ratio detector becomes misaligned after a few years of use. The symptoms may be loss of volume, fuzzy audio, and even squeals and buzzing. Many technicians use a VTVM or FETVM to do the job. However, a much better way is to use a sweep generator and scope.

The hook-up is very simple. Connect your scope across the volume control and then inject the sweep generator signal into the limiter if the receiver is using a discriminator or into the last IF amplifier if it has a ratio detector. Next, set your sweep generator to the receiver IF. A word of caution! Many receivers have an IF of 10.7 MHz and a bandpass of 200 kHz, *but not all*. In fact, some of the less expensive receivers have a bandpass of only 50 kHz. Therefore, when in doubt, be sure and check the manufacturer's service notes.

You now want to disable the local oscillator. To do this, simply place a jumper lead across the oscillator coil. Now, turn all equipment on and let it warm up for at least one-half hour. (An hour is better, if you have the time.) Your next step is to adjust your scope horizontal sweep frequency to 60 Hz. You will see a curve something close to the one shown in Figure 9-6, with the signal generator sweeping 200 kHz.

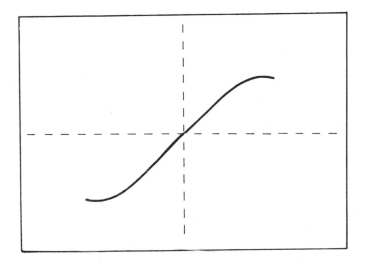

Figure 9-6: An example of a discriminator curve as seen on a scope

To adjust the discriminator, tune the slug in the secondary of the discriminator transformer until the curve appears to be equally divided on both sides of the vertical line on the face of your scope. Then adjust the primary slug until you have a good-looking S-shape curve at about a 45° angle, something like shown in Figure 9-6. Now, set the horizontal sweep frequency of your scope to 120 Hz and you should see a new pattern on the scope. It should look like the one shown in Figure 9-7.

For a perfect alignment, adjust the secondary coil slug until the two curves form as perfect an X as possible. In other words, place the crossover at exact center as shown in Figure 9-7.

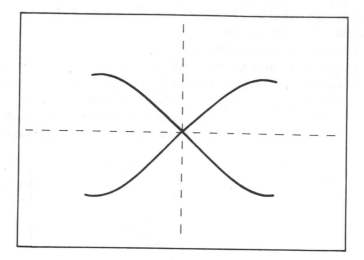

**Figure 9-7: Scope pattern showing 2 S-curves for
discriminator alignment**

An Easy Way to Align an FM Receiver RF Oscillator

To check an FM receiver to see if the RF oscillator needs alignment, tune to the top and bottom of the dial and see if either end has a loss
in gain. If you find a loss of gain, the oscillator should be realigned.

To begin, set your sweep generator to the FM band (anywhere
between 88 to 108 MHz) and adjust the sweep according to the manufacturers' recommended bandwidth. You'll find that this can be anywhere
from 50 kHz, for inexpensive receivers, to between 150 to 300 kHz on the
more expensive equipment.

Next, place a high impedance voltmeter (such as a VTVM, etc.)
across one side of the detector. However, do not short the oscillator coil
because this is the circuit you're trying to align. With the voltmeter
connected, tune the radio to 90 MHz and set your sweep input to the same
frequency. To align the low end of the dial, adjust the coils in both the RF
input circuit and oscillator, doing the RF input tank circuit first. You are
looking for a maximum voltage reading on the voltmeter.

Your next step is to set the generator and receiver to 107 MHz,
then adjust the trimmer capacitor of the RF input tank and oscillator,

again doing the RF input tank first and looking for a maximum voltage reading. Finally, put away your equipment because that's all there is to it.

Quick and Easy-to-Make AM Modulation Measurement

How often, when working on amplitude modulated equipment, have you needed a quick check to determine the percent of modulation? The following method is not only simple and easy but also has the advantage of not requiring you to tie up a scope, and it's even possible that it might not require any additional equipment. For example, if there already is an RF ammeter in the system, all you have to do is note the meter readings during modulation and refer to Table 9-1 to get the percentage of modulation.

ANTENNA CURRENT INCREASE WITH MODULATION	PERCENT OF MODULATION
22.5%	100%
18.5%	90%
15.5%	80%
11.5%	70%
8.6%	60%
6%	50%
4%	40%
2.2%	30%
1%	20%
0.25%	10%

Table 9-1: RF current increase vs percent of modulation

Here's how it works. Let's assume you're reading 500 mA with zero modulation. Next, you apply some unknown amount of modulation and now read 557.5 mA. Now, what you want to find is the percent of current increase. To do this, divide 57.5 mA (unit increase after modulation) by 500 mA (the original reading) and multiply the result by 100. The answer is 11.5%. Referring to Table 9-1, you can see that the percent of modulation is 70%. Now that you see how it works, try a few of your own measurements to prove to yourself how easy this system is.

How to Measure Percent of Modulation with a Scope

When using an oscilloscope to measure the percent of modulation, there are a couple of ways to observe the wave forms. For example, if you place a sample of the modulated carrier wave on the vertical plates and a sample of the modulating voltage on the horizontal plates, you should see a trapezoidal pattern similar to the one shown in Figure 9-8.

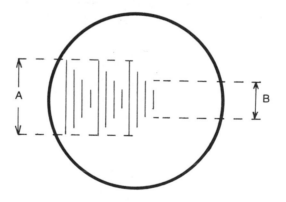

Figure 9-8: A percent of modulation measurement using a trapezoidal pattern

Then measure the height at (A) and also at (B), as shown in Figure 9-8. After you have made the two measurements of the pattern, simply use the percent of modulation formula, which is

$$\text{percent of modulation} = \frac{A - B}{A + B} \times 100$$

As an example, let's assume that you measure 3 inches for the large vertical (A) and 1/2 inch on the small vertical (B). Inserting your numbers into the formula, you get

$$\text{percent of modulation} = \frac{3 - 0.5}{3 + 0.5} \times 100 = 71.4\% \text{ modulation}$$

If your scope presentation is a sinusoidal pattern as shown in Figure 9-9, just measure (A) and (B) and use the same formula.

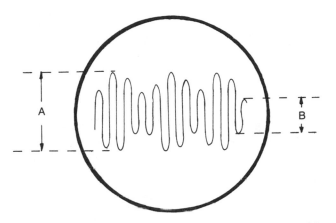

Figure 9-9: Measuring percent of modulation with a sinusoidal pattern

How to Calculate the Power Requirements of an Amplifier

Every amplifier must have a power supply. However, the question is how much power does the power supply have to deliver? For instance, many technicians build "home-brew" transmitters and need to know the power requirements before they start construction. The following methods will let you know what you're going to need with just a few simple calculations.

Let's assume that you want to determine the power requirements of an amplitude modulated class C amplifier that is going to produce 1,000 watts output power. There is a rule-of-thumb that will serve you well here. It is this: "Class C amplifiers are about 80% efficient, class B, about 50%, and class A, 25% efficient.'' Now at this point, all we know is the carrier output and the approximate efficiencies of the amplifiers. Therefore, it's best to start at the output and work back. Figure 9-10 is a simplified schematic that can be used to analyze the problem.

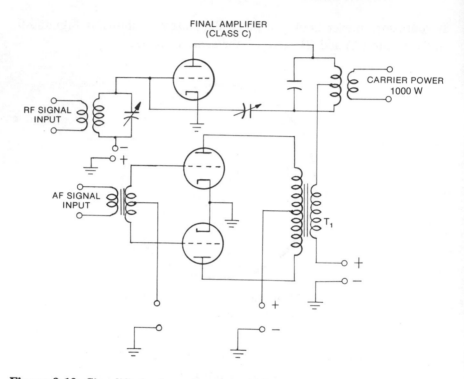

Figure 9-10: Simplified schematic of an AM modulated amplifier

Begin by calculating the power that is required for the modulated amplifier. This is easy because you know that the effiency is 80% and that the power is 1 kW. Simply use the formula

power in = power out / efficiency

and your answer is 1,250 watts. In other words, to supply this stage and be on the safe side, you should have a power supply capable of providing at least 1.5 kW.

To determine the power required by the modulator, let's start off by assuming that you want the modulator to be capable of modulating the amplifier at least 85%. The formula to calculate the power out of the modulator is

$$\text{power output of the modulator} = \frac{\text{percent of modulation} \times \text{power into the final amplifier}}{2}$$

In this case, the power output of the modulator is 451.5 watts, as you can see from the following calculations.

$$\frac{(0.85^2)\,(1250)}{2} = \frac{903}{2} = 451.5 \text{ watts}$$

You now know that the power input to the modulated stage (in at T1) is approximately 451.5 watts for a required 85% modulation. Finding the power supply requirements for the modulator is simple because you know the efficiency of the class B amplifier is about 50%.

$P_{in} = P_{out} / eff = 451.5 / 0.5 = 903$ watts

Therefore, your total power for the two power supplies would be about 2,153 watts, which adds up to a pretty hefty load.

How to Calculate Tube Power Dissipation in a Modulated Amplifier

Since all tubes, transistors, and other electronic components have a maximum power dissipation rating, you should know the power dissipation that you expect the amplifier to produce. This is easy to determine once you know the power input and power out of the amplifier. The formula is simply this: power dissipated = power in − power out. For example, the power that must be dissipated in the amplifier *without modulation*, in the case we've been discussing is

$P_{dis} = P_{in} - P_{out} = 1250 - 1000 = 250$ watts

But, this isn't the whole story because this tube also must dissipate power put into the amplifier by the modulator. The power from the modulator was 451.5 watts and the modulated amplifier is 80% efficient. Now, power out equals efficiency times power in, so 0.8 × 451.5 =

361.2 watts. Therefore, power dissipated equals 451.5 − 361.2 = 90.3 watts. The tube you select for the modulated amplifier must have a maximum power dissipation rating of 250 plus 90.3, which gives us a grand total of 340.5 watts.

Practical Troubleshooting
Techniques for Remote Control Systems

All of us who work with electronic equipment are constantly faced with servicing systems we've never seen before. This chapter concentrates on troubleshooting techniques that you will find very helpful when working on remote control units. It has many suggestions especially valuable if you're repairing a remote control system you're not familiar with.

What You Should Know About Remote Control Systems

In every case, you'll find that the remote control system consists of two main components—the transmitter and receiver. If the transmitter is designed for radio control of model aircraft and the like, it will more than likely use a Digital-Proportional system. Figure 10-1 shows a block diagram of this type of transmitter.

Remote control transmitters used with television receivers normally have a calculator type keyboard with buttons to turn the set on and off, switch channels, select channel identification (this shows up on the face of the CRT), and control the volume. Typically, you'll find anywhere between 8 to 15 buttons depending on how many functions are controlled. Also, in almost every case you will find that the output of the transmitter is in the ultrasonic frequency region, generally between 35

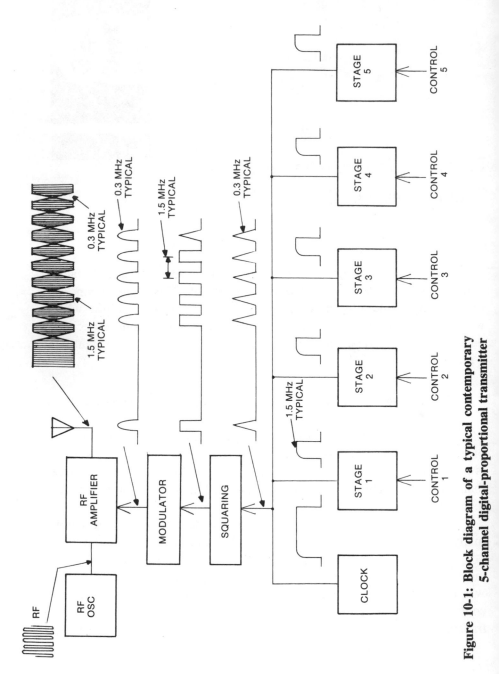

**Figure 10-1: Block diagram of a typical contemporary
5-channel digital-proportional transmitter**

and 45 kHz. In this type of system, the transmitter transmits a different frequency for each function. For instance, an 8-function transmitter must produce eight different frequencies.

As another example, a digital system produced by Magnavox generates fifteen ultrasonic control pulses. These pulses are 720 Hz apart. The receiving unit in the TV receiver counts the incoming frequency to decode and identify the function and then the logic section determines what function the signal controls.

First Steps for Troubleshooting TV Remote Control Units

A technician troubleshooting a strange piece of electronics equipment and a detective have a lot in common—they are both searching around the scene of the crime for clues. Regardless of what type of TV remote control system you are working on, there are some general troubleshooting procedures that can give you a clue to which systems are working. Of course, once these systems are eliminated it doesn't take too long to locate the bad component in the one that isn't working.

To begin with, if any function switch will produce the desired result it means that the transmitter battery, both transmitter and receiver transducers, transmitter oscillator and receiver oscillator (if there is one) are all functioning properly. Using the circuit in Figure 10-2 as an example, you can see why this is true in the case of a transmitter unit.

Notice that this system (as well as almost all of the others) uses the same oscillator and transducer for every function. Therefore, if you can actuate any function, these components must be operating properly. For the moment, let's assume that you can't get anything to respond. In this case, the easiest first step to determining if the problem is in the transmitter or the receiver is to substitute a known-to-be-good transmitter. But if another transmitter is not available, an easy place to start is to check the transmitter battery. Again, the best thing is to insert a known-to-be-good component, in this case the battery. However, if you don't have a new battery, the next best thing is to use a voltmeter. Incidentally, be sure and leave the battery in the transmitter when you make the following measurements because it is necessary to check a battery *under load* in order to get an accurate voltage reading.

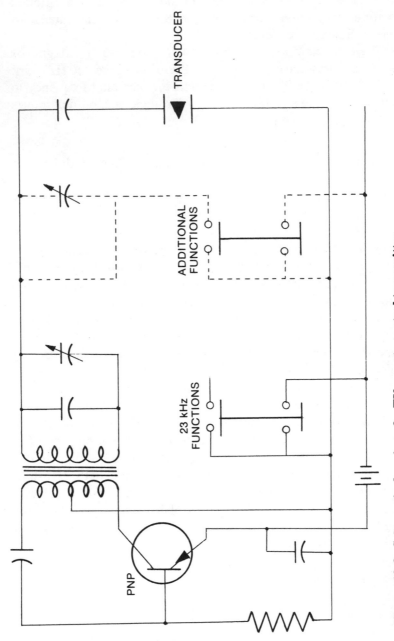

Figure 10-2: Schematic drawing of a TV remote control transmitter

How to Check a Remote Control Transmitter Battery

When you make a remote control transmitter battery check, it's better to press each button one by one and observe your voltmeter reading. If everything is okay, you should see a reading somewhere between 7 and 8 volts. A good rule of thumb when checking any circuit like this one is this: A maximum voltage (the full battery voltage) indicates an open circuit and zero voltage indicates a short circuit (assuming that the battery is good). Should you see either of these conditions as you press the buttons, it means that the transmitter is defective.

Successful In-Circuit Transducer Testing

As already explained, the output of the transmitter remote control frequently is an ultrasonic frequency of somewhere between 35 and 45 kHz depending on which button you press. You can use either a high impedance voltmeter or scope to check for an output. The output should be several hundred volts each time you press a button. Low, or no voltage, when you press any of the buttons indicates a trouble in that circuit.

When using a scope to check the output you may not even have to make connections. Simply turn the vertical gain of your scope up to maximum and hold the remote control unit close to the scope's vertical input. You should see a signal on the scope each time you press a button, providing the circuit is working and that the transducer, etc., are all right.

To check the transducer, in case you don't get any output, measure the voltage at the transducer input. If you measure a fairly high voltage (several hundred volts peak-to-peak), normally you can assume that the transducer is defective. However, if you find that the voltage at the transducer input is zero or very low when you press one of the buttons, you'll have to check on farther back. Generally, you'll find a capacitor in series with the transducer. When you measure the voltage across this capacitor, it should be about the same voltage as normally found across the transducer. If it's high—say about 600 to 1,000 volts —you have a defective capacitor. On the other hand, if the voltage on the capacitor and transducer is zero, your next step is to check the transistor in the circuit because this is more than likely your problem.

Practical TV Remote Control Transmitter Alignment

All you need to align many of the TV remote control transmitters is a signal generator with the correct frequency range and a scope. Or, in place of the signal generator, you can use another remote control unit if you have one.

Usually, you'll find that each of the resonant circuits (there's one for each function) in the transmitter has separate trimmer capacitors for alignment purposes. A simple way to make the alignment is to place

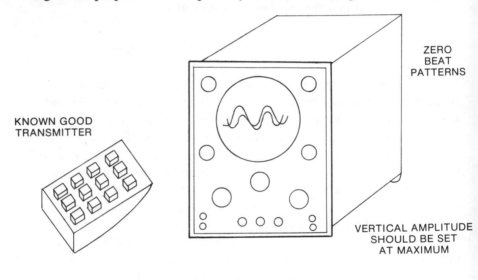

ZERO
BEAT
PATTERNS

KNOWN GOOD
TRANSMITTER

VERTICAL AMPLITUDE
SHOULD BE SET
AT MAXIMUM

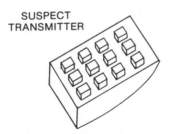

SUSPECT
TRANSMITTER

**Figure 10-3: Setting up for a remote control transmitter
            alignment**

either the signal generator or another transmitter at the scope's vertical input terminals. Next, place the transmitter that you are working on very near the scope's vertical input terminals. Now, with both transmitters operating and the scope's vertical gain set to maximum, you adjust the trimmer until you see a zero beat on the scope. Figure 10-3 shows the setup. Normally, it's best to start at the highest frequency and work down button by button.

Guidelines for Working with Remote Control Receivers

There are several designs now being used in remote control receivers. In many of them, you'll find a receiver that uses a transducer exactly the same as used in the transmitter. Also, in some cases there are electric motors and step relays that you may have to check and adjust. However, in other systems there are **IC**'s that work like a computer. Figure 10-4 shows a block diagram of one **LSI** chip used by Magnavox in one of their TV receiver systems.

There are feedback networks that let the system find the desired station and then the tuner is locked to an internal oscillator by an afc loop. The tuners are solid state and aren't too hard to replace. Furthermore, after you've replaced the tuner, all that is left to do is align the IF amplifiers. Typically, sets of this type have no fine tuning adjustments or any other moving parts.

At times, you may be called on to service radio controlled systems. Therefore, you should have some idea of what to expect. So, let's look at a remote control receiver that is widely used in radio controlled equipment. Figure 10-5 shows the system layout and the signals to be expected.

The receiver is used in conjunction with the digital-proportional transmitter shown in Figure 10-1. It is typical of modern **IC** decoders found in many types of radio controlled model boats, airplanes, cars, and so on.

How to Make a Remote Controlled Receiver Alignment

Generally, the alignment of these receivers is very easy. You can

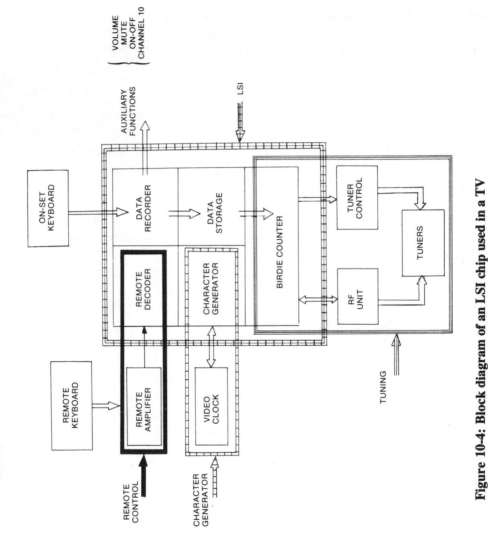

Figure 10-4: Block diagram of an LSI chip used in a TV remote control receiver

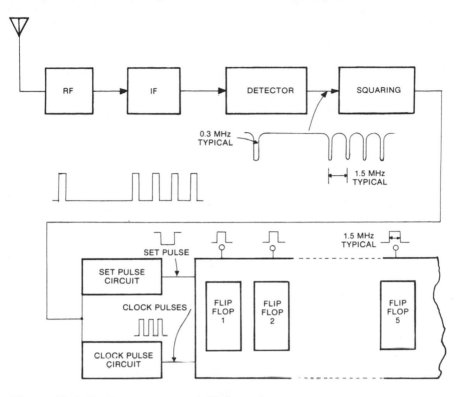

Figure 10-5: Radio remote controlled receiver system

use either a transmitter or a signal generator for a signal source. Place a high impedance voltmeter across the common emitter resistor in each stage—one at a time—as you actuate each function. Now, what you want is to read maximum voltage as you tune each input coil. It should be pointed out that voltage in this case normally isn't very much so you will have to use a low voltage range while making the adjustments. Figure 10-6 shows the meter connections for this type of relay controlled circuit.

What to Do If a Remote Relay Will Not Actuate

If one of the relays won't actuate yet the rest of the relays all perform normally, the trouble must be in the driver circuit or the relay itself. A simple ohmmeter check will tell you if it's the relay. If the relay

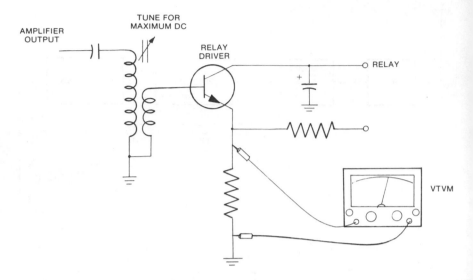

**Figure 10-6: Setup for tuning a relay controlled
transistor circuit**

is good, go to the base of the driver transistor and inject a signal. Assuming the relay is actuating a motor, both the relay and motor should operate. If the relay does close, the trouble is in the feed circuit of the transistor, which could be the coupling capacitor or tuned transformer. But if the relay doesn't close when you inject the signal at the transistor base, more then likely you have a bad transistor although it could be a defective output filter capacitor. See Figure 10-7.

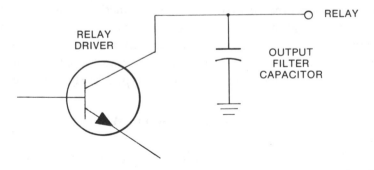

Figure 10-7: Relay driver output filter capacitor

How to Troubleshoot a Stepper Relay

You'll find stepper relays used in many remote control systems. For instance, in TV it is commonly used to increase volume gradually. Each time you press the button on the transmitter it should actuate the relay and move one step. To troubleshoot the circuit, follow exactly the same procedure as explained in the last section. In other words, place a signal at the input of the driver relay. If the relay moves one step, start checking back from the transistor input. If not, check the transistor and relay.

Troubleshooting a Remote Control Motor

Remote control motors are designed to do different things depending on where and why they are used. Normally, for a certain channel button the motor should start running and continue running until it reaches the correct station without you having to hold the button down. In this type of operation, if the motor quits as soon as you let up on the button it is an indication of a malfunction. Now, because the motor does operate when you press a button it means that all circuits are functioning properly except the motor holding switch which is found at the motor itself. Therefore, start at the relay output and check component by component up to and including the motor.

There are several other troubles that you may encounter. For instance, the motor may run past the wanted station or stop at unwanted channels. If it runs past the wanted station, of course the motor is working properly so the trouble must be in the control circuits. In this case, check the wafer switch that you'll find on many sets. It probably has a loose wire, or possibly it is broken or cracked. If the set keeps stopping at channels with no station, check the hold switch at the motor and the wafer switch that handles the control voltage to the hold switch because these are common breakdown points.

Troubleshooting Phase Shift Synchronous Motors

Phase shift synchronous motors can run backward or forward. The direction depends on the applied AC voltage. You'll probably find that the phase shifting circuit is a simple resistor and capacitor in series with one of the motor windings. If the motor will go one way but not the other, you know that the motor is good and all the circuits for that function are good. One quick way to check out the transistor in the nonoperative circuit is to place a jumper wire from the good output transistor collector to the inoperative transistor collector. If the relay closes and the motor starts to turn, your problem is probably the transistor. However, it may be in the components in the transistor input circuit, so, if possible, inject a signal to be sure before you pull the transistor.

Practical Microwave
Troubleshooting

Microwave troubleshooting is a major part of the technician's work in a broad range of applications. Furthermore, the overall trend is toward microwaves. This chapter will show you that working with microwave equipment is fairly simple once you know the rules of the game.

How to Make and Use a Standing Wave Indicator

On the bench or in the field during an actual installation or equipment test, a standing wave ratio (SWR) measurement can be very important if you are working in weak signal areas, or any other time signal strength is of prime importance. Furthermore, an SWR measurement can provide you with considerable information about the equipment you are working on. For example, you might want to measure the unknown resistance value of a dummy load. One way to do this is to use an SWR measuring device and slotted line.

Many technicians don't have a commercial slotted line and SWR meter around the shop because of the cost. However, an RF voltmeter and a pair of Lecher wires will do a pretty good job. See Figure 11-1, which shows how to construct a Lecher wire system.

The line impedance (Z_0) should match the source impedance for best results. To determine this value, use the formula $Z_0 = 176 \log S / R$, where S is the center-to-center spacing and R is the wire radius. Now,

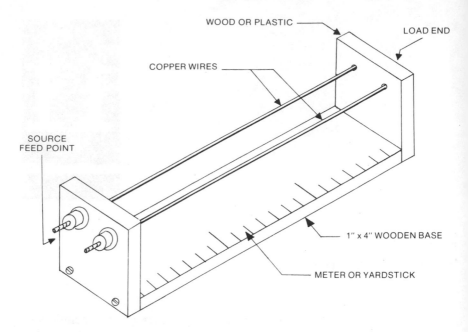

Figure 11-1: Lecher wires

let's assume that your source impedance is 50 ohms and that you have designed the Lecher wires to have the same impedance. If you place an unknown dummy load at the load end and find that the voltage across the two wires is the same everywhere, it means that the unknown load is also 50 ohms. However, if you should find there is a standing wave, the first step is to determine whether the load impedance is larger or less than the line impedance. The way to do this is to measure the voltage at the load and determine if it is a maximum or minimum at that point. If it is maximum, the load impedance is greater than the line impedance. If it is minimum, the load impedance is less.

Let's keep it simple and say that you find a maximum voltage at the load and the voltage standing wave ratio (VSWR) is 2:1. Now, all that is left to do is to multiply the VSWR (2) by the line impedance (50). The load impedance is 100 ohms. But be careful because if the voltage had been a minimum at the load, you would have had to divide the line impedance by the VSWR. Also, be sure and remember that this only works for pure resistance loads.

An Easy Way to Calculate Power Out Using SWR

Now that we know the value of the dummy load, let's go another step and show how SWR can be used to calculate the power dissipated. To do this, you will need the formula

$$P_L = P_T \ \frac{4 \ SWR}{(SWR + 1)^2}$$

where P_L is power to the load and P_T is power out of the signal source.

If your signal source output is 50 watts and SWR is 2:1, you will wind up with 44.44 watts. Your calculations would look like this:

$$P_L = 50 \ \frac{(4) \ (2)}{(2 + 1)^2} = 44.44 \ watts$$

This formula will really come in handy when working with terminating resistors and various other type dummy loads.

Safeguards for Microwave Measurements

A factory-built slotted line is far more accurate than the simple Lecher wires and is one of the most important measuring instruments found in a microwave equipment repair shop. Figure 11-2 shows a coaxial type in wide use.

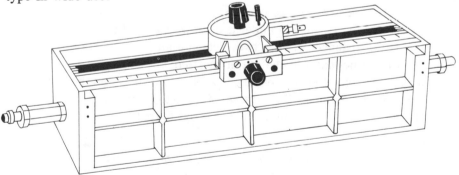

Figure 11-2: Coaxial slotted line

Operation of the slotted line is simple and straightforward. A probe is mounted on a carriage that slides along the outside of the section of coaxial line. The probe extends into the slot and the depth of penetration is adjustable. This probe is used to sample the RF voltage, with its output usually being fed through a detector to a standing wave meter. This simple setup can measure the standing wave or wavelength of the signal frequency which, in turn, can provide a wealth of other information about the equipment under test. However, there are some dangers if certain rules are not observed.

How to Avoid Erroneous Readings When
Troubleshooting with a Slotted Line

Probably one of the most common causes of error when using a slotted line to measure standing waves is too much probe penetration. What happens in this case is that you will read a lower standing wave ratio than the true SWR. Now, considering that the SWR is an indication of system performance, it becomes obvious that it would be possible to think that the system is within specs when it really isn't. The rule is this: Keep your probe penetration as small as possible to get the most accurate reading.

A Simple Harmonic Testing Technique

It is possible to adjust the probe so that it will respond to a harmonic of the frequency under test; however, it is easy to check. Just measure the distance between two voltage minima on the slotted line and this will be one-half wavelength of the frequency the probe is tuned to. Usually it is a good idea to place a low pass filter between the source and the slotted line to reduce the harmonics down to a negligible value. This is particularly true if the source has a coaxial output.

Troubleshooting a Signal Source for
Frequency Modulation

To save yourself a headache, you should be sure to check and see if the source is providing a stable output. Any frequency variations on the output of the signal source is called *frequency modulation*. To check for this trouble, short the output of the slotted line and see if the minimum point of the standing wave is found at the same slide setting. Move the slide back and forth a little ways and then check the minimum. If it's the same slide setting after three or four tries, you're okay. If not, you may have poor oscillator voltage regulation in the source you are using.

How to Make and Mount a Waveguide Feed System

Every microwave technician sooner or later has to connect a coaxial cable to the waveguide either to inject or remove an rf signal. The easiest way is to use commercial coax-to-waveguide feed systems. However, if you are working on experimental systems or on a low budget, it can be done as follows: The first step is to drill a hole at either A or B as shown in Figure 11-3 and mount one of the coaxial fittings shown in Figure 11-4.

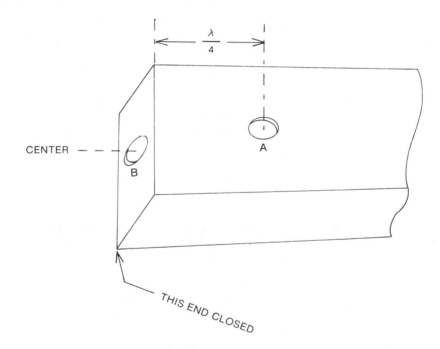

Figure 11-3: Where to drill into a waveguide

(A)

(B)

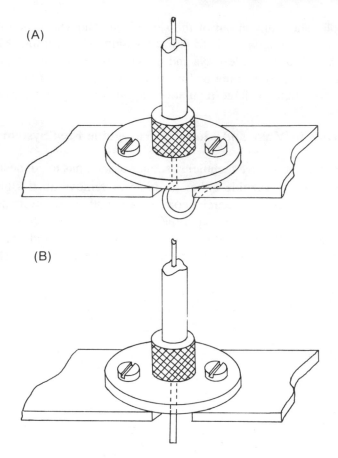

Figure 11-4: Coaxial coupling to a waveguide

The (A) drawing is coupling through use of the magnetic fields and may be placed at any position as long as it links with the magnetic lines of force. The easiest way to do this type of coupling is to strip the center conductor of the coax and then bend and solder the loop as shown in drawing (A). If you excite the coax, you should see an output at the open end of the waveguide. This can be measured by an RF power meter or a simple field strength meter. If you don't find an output, rotate the loop and it should do the trick because, as you will remember, the position of any coil affects the degree of coupling.

The (B) drawing in Figure 11-4 is a probe penetration type and all that is required is that you extend the center conductor of the coaxial cable into the waveguide and the outer conductor be terminated at the wall of the guide. But, be sure to notice that the hole is one-quarter wavelength from the closed end.

An Easy Way to Change the Impedance
of a Waveguide

All of us know that it is possible to tune a transmission line by inserting a capacitor or a coil in the line. However, when you are working with a waveguide, at first it would seem that this is an impossible task because there is no center conductor as in a coaxial cable. Therefore, the question is, "How can you increase the capacitance if you want to tune a waveguide?" Well, it may be hard to believe but in the early days of

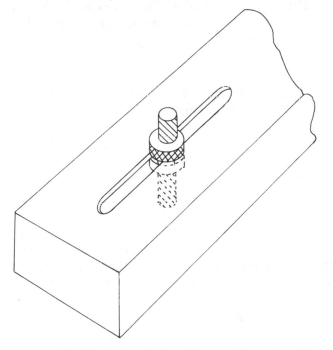

**Figure 11-5: Changing the impedance of a waveguide
transmission line**

space communications some technicians (and engineers) actually placed C-clamps on the waveguides and permanently bent the waveguide walls inward, which effectively increased the line capacitance and tuned the transmission line. Of course, this permanently damaged the waveguide, but I've seen it done more than once. The tragedy of it all is that it doesn't have to be done this way. Figure 11-5 shows an easy way to tune a waveguide transmission line without damaging it.

The tuning screw shown is capacitive, but if the screw is long enough it becomes a series resonant circuit if the penetration is one-quarter wavelength. If you continue to penetrate deeper, it will begin to appear inductive. Furthermore, if the diameter of the screw is decreased or if it is moved away from the center of the guide, the susceptance of the equivalent capacitor or inductor will decrease.

Now, all you need to do is slot the waveguide near the source of reflected waves and make the screw adjustable as shown and you will have a considerable range of tuning. Doing this will make it possible to bring the standing wave ratio of a short line right down to almost one-to-one, with a little patience and fine tuning.

Pitfalls to Be Avoided When Using the
Smith Chart During Troubleshooting

Most examples of using a Smith Chart assume a lossless transmission line. However, on a practical line the SWR will decrease as the length of the line is increased. Now, this means that the input impedance will vary even though the line may be terminated in an accurately matched load. Therefore, when you make SWR measurements at the input end of the line, they are not strictly true for the entire system —which means that you really don't know what the mis-match is between the load and transmission line. The error will be small if the transmission line is short and has little loss, which most coax does. But, the error can get out of hand if you are working with a large SWR and the line loss is high. Some of the coaxial cables sold in various electronic supply houses are government rejects and have considerable loss, so watch out for these bargains because they may not be bargains after all.

Here is an example of how to find true values of SWR and load impedance if you know the total line loss. In this case, the cable length is about 150 feet of small 50 ohm coax operating at 28 MHz, with a line loss of 3 dB. The input impedance of the 50 ohm terminated line with a 3 dB loss was measured to be approximately 100 ohms with negligible reactance. Now, this measured value was normalized and plotted on a Smith Chart and the SWR circle and radial line drawn, as shown in Figure 11-6.

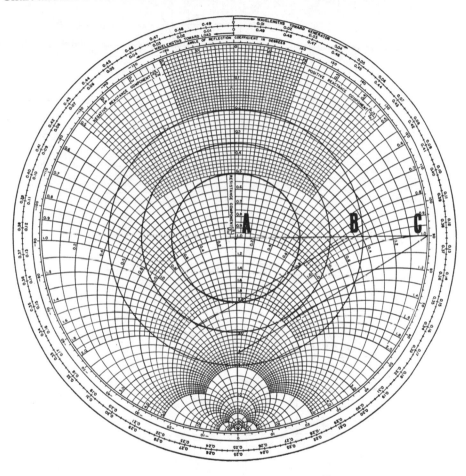

Figure 11-6: Correcting for losses in a transmission line

Notice, the measured SWR is 2:1. Now, we can use another chart such as the one shown in Figure 11-7. This chart shows the relationship between measured SWR with a shorted output and the loss in dB meas-

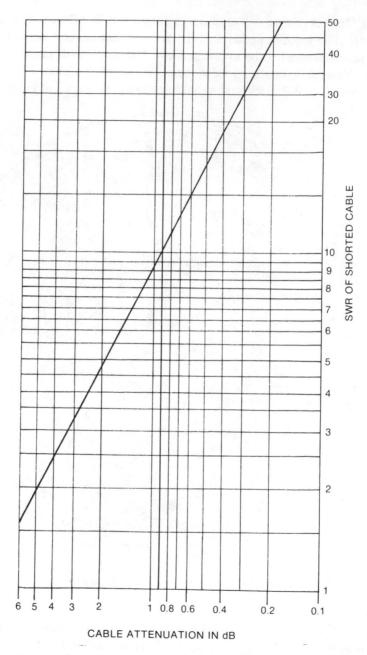

Figure 11-7: Cable attenuation vs. SWR

ured when the line is properly terminated. Some Smith Charts have this graph printed right on the chart.

The graph in Figure 11-7 shows that a line having a loss of 3 dB, terminated in its characteristic impedance, will have an SWR of 3 when shorted. This SWR circle is also drawn on the Smith Chart in Figure 11-6. Next, a radial, ABC, is drawn perpendicular to the vertical resistance axis. Now, draw a perfectly parallel line to line B to 3 and, when this parallel line passes through the resistance axis, you will read the true SWR (4.8). Take note of the fact that the ratio of measured impedance to line impedance shows an SWR of only 2:1. Finally, the true value of load impedance is 4.8 × 50 = 240 ohms. Incidentally, the next time someone tells you he has an SWR of 1:1, be a bit cautious about believing him unless you know for a fact that the transmission line is short and has very little loss.

How to Convert VSWR to Return Loss

Some manufacturers prefer to spec the match of amplifiers and transmission lines in terms of return loss (RL) instead of VSWR. The term *return loss* is frequently used in cable TV where amplifiers are spaced every 0.5 miles over several miles. Return loss usually is expressed in dBs and can easily be changed back to VSWR or vice versa by use of the formula

$$dB = 20 \log 1 / \Gamma$$

where Γ is reflection coefficient. For example, a return loss of 20 dB is equal to a VSWR of 1.22 to 1.

When reading amplifier specs, remember that the more dBs or return loss, the better the match or, on the other hand, the closer to 1:1 the SWR, the better the match. VSWR and return loss give the same qualitative information once you are familiar with the numbers.

How to Find Peak Pulsed Power

Pulsed modulation is in very wide use in microwave test equipment, and it is frequently necessary to calculate the peak power of these pulsed signals. Most power measuring devices found in a shop respond to average power. However, it's fairly simple to calculate the peak power from the average reading. The formula you will need is

$$\mathbf{P_p} = \mathbf{P_a} \; 1 \, / \, (\mathbf{r}) \, (\mathbf{fr})$$

where $\mathbf{P_p}$ is peak power in watts, $\mathbf{r}$ is pulse length, $\mathbf{P_a}$ is average power reading, and $\mathbf{fr}$ is pulses per second. You will also hear the quantity $(\mathbf{r}) \, (\mathbf{fr})$ called the *duty cycle*. A word of caution! The formula only applies to pure pulse modulation and not to composite video signals such as used in television broadcasting.

Let's assume you are working with a microwave source that is pulsed at an audio rate of 800 times a second and each pulse is one microsecond (10^{-6}) in duration. Next, let's say that your average responding meter reads 0.8 watts. Inserting these values into the formula gives

$$\mathbf{P_p} = \mathbf{P_a} \; 1 \, / \, (\mathbf{r}) \, (\mathbf{fr}) = (0.8) \; 1/ \, (10^{-6}) \, (800) = 1{,}000 \text{ watts}$$

This is the maximum power that will be delivered to your load over the period of the pulse. However, be sure all your equipment is as perfectly matched as possible because VSWR causes reflected power and will cause your readings and calculations to be in error. Also, it is best to use a thermistor type power meter during measurements because the internal resistance of certain other types will vary during pulse modulation.

Troubleshooting Electric Motors
and Three-Phase Systems

A big headache to many electronic technicians is the area of electric motors. Electric motors are not hard to select, service, or repair if the tricks and techniques that make the job easy are known. This chapter will show you how to whip some of the toughest motor problems and present many work-saving ideas that you can use in the shop.

Also, without a doubt the three-phase power distribution system is the most widely used in this country. Therefore, it is important for every technician to be able to make the simple but effective measurements described in the following pages.

How to Solve Electric Motor Problems

You can use a motor designed to operate on 110, 220, or 440 volts and get satisfactory service, provided the voltage is fairly stable at the motor terminals. Let's digress for a moment and ask, "Why would anyone want 220 instead of 110 volts?" Answer: "Because power equals volts times amps., the current flow at 220 is one-half of what it is at 110 volts." This is a good point to remember because it means you might be able to use a smaller wire and, therefore, save some money. Furthermore, the higher voltage motors are more efficient, providing even more cost cutting.

Now, back to the problem of a constant voltage. To assure a constant voltage at the motor terminals, you should use either the 220 volt

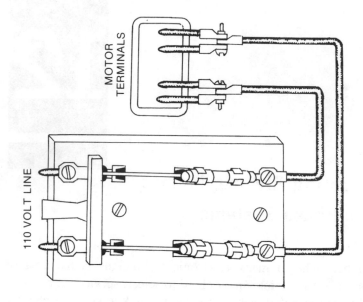

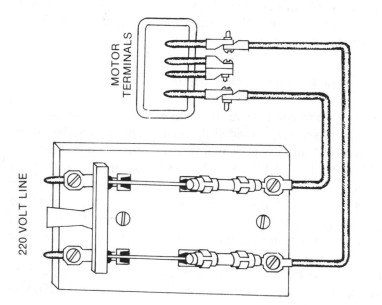

Figure 12-1: How to connect a motor for either 110 volt or 220 volt operation

or 440 volt and a large size wire. The larger wire is very important if you expect any overload. For example, a motor that draws 4 amps under normal load may draw 7 or 8 amps on start-up.

Another related problem that many of us have run into is the problem of low line voltage. This can be really troublesome in the case of electronic cooling systems equipment. For example, cooling fans and the like are frequently wired for 115 volts. It is important that this voltage does not drop much below 100 volts because low voltage will cause a *rapid* drop in horsepower. To give you some idea of what will happen, a 110 volt motor operated at 95 volts will produce only 75 percent as much horsepower as it would at 110 volts. In the case of cooling equipment, this could be disastrous, so get on it immediately. Generally, the first step in troubleshooting is to check for low voltage at the motor terminals. If you find that this is the problem, look for a loose connection, especially if the equipment is mounted in a trailer or truck. To save yourself a lot of trouble when working on motor leads, always use good solid mechanical clamps and then solder all connections. To avoid the expensive and time-consuming job of tearing out the first installation, if the motor is within 20 or 30 feet of the power source, don't use under 14 gauge wire. However, if the distance is over 50 feet the number should be 11 or 12 gauge.

Except for some special motors and reversible ones, many single-phase motors are designed to operate on two different voltages such as 110 or 220. Figure 12-1 shows two common methods you will find being used for the wiring connections. As explained earlier in this chapter, it's always better to use the 220 hookup if you can.

How to Select an Electric Motor Fuse

What do you do when fusing a heavily loaded electric motor? Well, first of all don't use a common house plug fuse. Why? Because you would have to use a fuse that is rated at the starting load of the motor and if the motor draws this extra current long enough, it may burn out the windings. There are approved fuses especially made for electric motors, available at your local electrical supply house. Suggested maximum running fuse ampere ratings for single-phase motors from one-half to two horsepower are listed in Table 12-1. All maximum fuse ratings are based

HORSE-POWER	SINGLE PHASE 115 V		SINGLE PHASE 230 V	
	MOTOR RATED CURRENT IN AMPS	MAXIMUM % ABOVE THE RATED CURRENT	MOTOR RATED CURRENT IN AMPS	MAXIMUM % ABOVE THE RATED CURRENT
1/2	7.4	74	3.7	61.6
3/4	10.2	68	5.1	63
1	13	15	1.5	65
1-1/2	18.4	73.6	9.2	76
2	24	80	12	80

Table 12-1: Suggested maximum fuse ratings

on a constant load and are the the maximum percent that you should use above the motor's *rated* current.

What You Should Know About Motor Efficiency

All theoretical books will tell you that an electric motor will produce one horsepower for each 746 watts of electrical energy it consumes. In other words, you should measure 6.782 amperes of current on a 110 volt, one horsepower motor. Don't you believe it! At least not when you are working with small motors. In practice, small motors (1/6 horsepower) that you'll find in many applications are only 40 to 60 percent efficient. However, as the horsepower increases, the efficiency increases. For example, one to two horsepower motors normally have efficiencies between 75 to 80 percent. Under ordinary circumstances, the larger the motor the better the efficiency and it can be up around 98 percent on the larger motors. To put it another way, efficiency can be a major factor in determining how much current a motor will draw. How to determine the amount of current is explained in the next section.

Useful Conversion Formulas for Electric Motors

We've all heard the statement, "When in doubt, get a bigger hammer." This used to be true when choosing an electric motor; that is, when in doubt, get a bigger motor. But we don't do that anymore. When you are determining a motor's rating, it is always better to choose a size that will be as close as possible to its rated capacity. If you don't, it results in low efficiency and with induction motors, also in a lower power factor.

To find the current (I) that an electric motor will draw, you must know several things, including horsepower (HP), voltage (E), efficiency (eff), and the power factor (pf). But, where do you get all this information? Easy. Just use your voltmeter, ammeter, wattmeter, and the table of formulas shown in Table 12-2. The various measurements are explained in the following sections.

TO FIND	DC	AC	
		SINGLE PHASE	THREE PHASE
CURRENT VALUE WHEN HP IS KNOWN	$\dfrac{HP \times 746}{E \times eff}$	$\dfrac{HP \times 746}{E \times eff \times pf}$	$\dfrac{HP \times 746}{E \times 1.73 \times eff \times pf}$
CURRENT VALUE WHEN kW IS KNOWN	$\dfrac{kW \times 1000}{1000}$	$\dfrac{kW \times 1000}{E \times pf}$	$\dfrac{kW \times 1000}{E \times 1.73 \times pf}$
CURRENT WHEN kVA IS KNOWN		$\dfrac{kVA \times 1000}{E \times pf}$	$\dfrac{kVA \times 1000}{E \times 1.73}$
KILOWATTS	$\dfrac{I \times E}{1000}$	$I \times E \times pf$	$I \times E \times 1.73 \times pf$
kVA (APPARENT POWER)		$\dfrac{I \times E}{1000}$	$I \times E \times 1.73$
pf (POWER FACTOR)		$\dfrac{kW \times 1000}{I \times E}$	$\dfrac{kW \times 1000}{I \times E \times 1.73}$
HP (OUTPUT)	$\dfrac{I \times E \times eff}{746}$	$\dfrac{I \times E \times eff \times pf}{746}$	$\dfrac{I \times E \times pf}{746}$

Table 12-2: Electric motor conversion formulas

How to Measure Power in a Three-phase Circuit

You can measure the power in a three-phase circuit with a poly-phase wattmeter or, if you don't happen to have a polyphase wattmeter, you can do it with two single-phase wattmeters connected as shown in Figure 12-2.

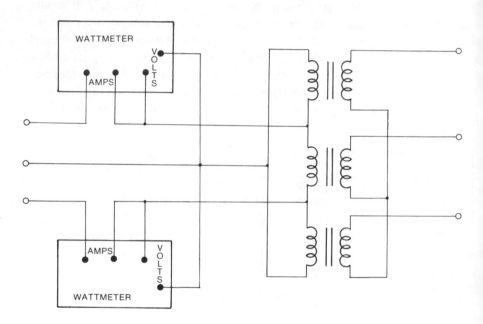

**Figure 12-2:   How to connect 2 single-phase wattmeters
in a 3-phase circuit**

After you have made the connections as shown in Figure 12-2, and providing the power factor is better than 50 percent (which it normally is), the true power of the system will be the sum of the two readings. If you should be working on a system with a power factor of exactly 50 percent, you will read zero on one of the wattmeters. And, if the power factor is less than 50 percent, the same meter will read negatively. What you must do in this case is reverse the connections and take the difference between the two meters to find your true power.

Practical Electric Motor Efficiency Measurements

Many technicians have no idea how to check the efficiency of an electric motor, but the secret is very simple. All you have to do is measure the power into the motor with a wattmeter and divide the reading into the horsepower output. For example, let's say that you have a motor rated to deliver 1/6 horsepower to a load and you measure 240 watts on your wattmeter as power in. Now, if you remember to convert *both* output and input to the same units of measure and that one horsepower equals 746 watts, the next step is a breeze. Here's all you have to do:

$$\text{Efficiency (\%)} = \frac{\text{power out}}{\text{power in}} \times 100$$

$$\text{Efficiency (\%)} = \frac{(1/6)\ (746)}{240} \times 100 = 51.5\%$$

When Can You Disregard Power Factor?

A low power factor in an induction motor ordinarily does not affect its operation or efficiency. But it does mean there will be more current drawn than would be if the power factor were equal to one. Generally speaking, except for the extra current that must be considered when choosing power lines, the power factor of motors is of little importance if you are using commercial power. Therefore, in most cases you don't have to include the power factor when using the formulas in Table 12-2. However, you may have to consider that some power companies charge extra for a low power factor if the customer is a fairly large consumer of energy.

A Handy Guide for Checking Full Load Motor Currents

The technician on the job needs all the help he can get while troubleshooting. Table 12-3 is a handy list that will make troubleshooting and installing electric motors a lot easier. It includes the currents you should expect to measure when working with the motors listed. But the values given are average, and therefore, you probably will find variations of about 10 percent in actual measurements.

H.P. OF MOTOR	DC MOTORS		SINGLE-PHASE MOTORS		AC MOTORS												
					SQUIRREL—CAGE MOTORS						SLIP-RING INDUCTION MOTORS						
					TWO-PHASE			THREE-PHASE			TWO-PHASE			THREE-PHASE			
	115V	230V	110V	220V	110V	220V	440V	110V	220V	440V	110V	220V	440V	110V	220V	440V
1/4	—	—	4.8	2.4	—	—	—	—	—	—	—	—	—	—	—	—
1/2	4.5	2.3	7	3.5	4.3	2.2	1.1	5.0	2.5	1.3	—	—	—	—	—	—
3/4	6.5	3.3	9.4	4.7	4.7	2.4	1.2	5.4	2.8	1.4	6.2	3.1	1.6	7.2	3.6	1.8
1	8.4	4.2	11	5.5	5.7	2.9	1.4	6.6	3.3	1.7	6.7	3.4	1.7	7.8	3.9	2.0

Table 12-3: Table of approximate full load electric motor currents

Tips for Troubleshooting and Servicing Electric Motors

Because it is frequently necessary to pull a complete unit out of its case to check and clean a motor, it isn't unusual to find a motor in some equipment that has not been looked at for several years. Some technicians will tell you that they follow the rule: If it's working, leave it alone. If you follow their advice, you're heading for trouble. For one reason, if you don't check the brushes and their holders occasionally, you may end up doing a major motor overhaul. You should periodically inspect the brushes and their holders to see that the brushes have not worn down so much that the brush holders are riding on the commutator. Be sure that the brushes are the same length and that the brush holders have at least 1/8 inch clearance above the commutator. If not, replace the brushes.

If you have to replace the brushes, an old trick is to use sandpaper to form the full face of the brushes to the commutator. Place the sandpaper between the commutator and brush, then turn the commutator back and forth until you get a good firm fit. Incidentally, while we are discussing brushes it's a good time to point out that they come in various degrees of hardness. It is very important that you use the manufacturer's recommended type. Otherwise, you may have rapid brush wear or, on the other hand, excessive commutator wear.

After you have set the brushes, it's best to use compressed air to blow all debris off and out of the motor. However, if you have to dismantle it for cleaning, to replace the brushes, or during troubleshooting, it is usually better to clean it carefully with cleaning fluid both before you dismantle it and after you put it back together. *Warning! Be sure that the cleaning fluid you use is not harmful to the electrical insulation material and other parts of the motor.* Also, always do your cleaning in a well ventilated and fireproof area.

During inspection, you may find that the commutator copper bars have worn down below the insulation. If this happens, hold a piece of coarse sandpaper (#1-1/2 or 2) against the commutator with a flat block of wood. What you want is to cut the insulation down to the level of the copper bars. If the commutator is badly burned or worn, it may be necessary to take it to a machine shop and have a small amount of metal taken off with a lathe.

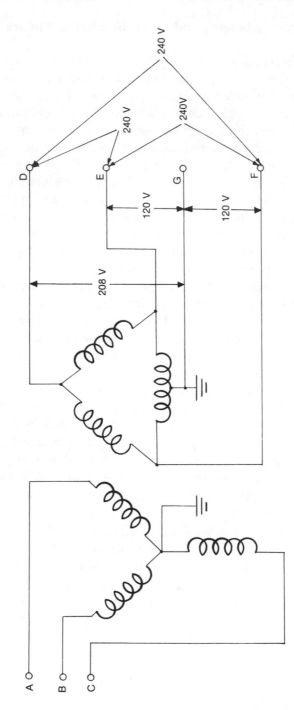

Figure 12-3: Typical output voltages of a wye-delta transformer with a split-phase winding

A Guide to Checking a Wye-Delta Transformer

Generally, a building or electronics equipment with both three-phase and single-phase loads will be using wye-delta transformers to supply the required energy. A schematic diagram with the typical secondary voltages you should measure is shown in Figure 12-3.

Suppose that you are working on a fairly large transmitter. You will find the 240 and 120 volt single-phase outputs providing power to the water pumps, blowers, and filament transformers. The 240 volt single-phase loads are connected at D-E, E-F, and D-F. The 120 volt single-phase loads will be between E and F to ground and the three-phase loads at D-E-F. More than likely, you won't find the 208 volt connection in use except where the 120 volt circuits are carrying much heavier loads than the three-phase circuits. In this case, the 208 volt connection may be furnishing power to a 220 volt single-phase load. The reason this is done is to reduce the currents in the split-phase legs E and F. It isn't unusual to find that three single-phase transformers have been used in the installation. The wiring diagram for this setup is shown in Figure 12-4.

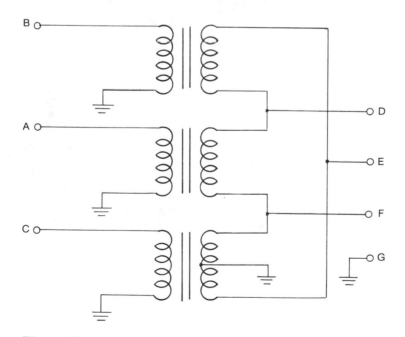

Figure 12-4: 3 single-phase transformers being used in a 3-phase power supply

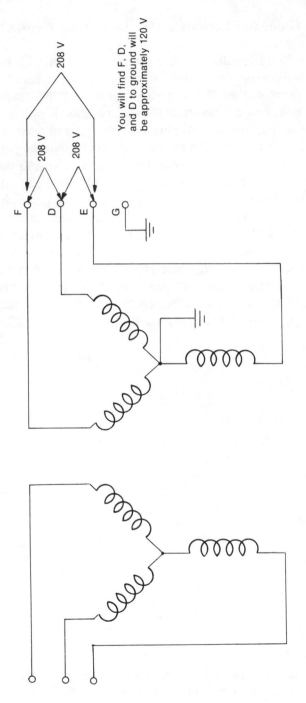

208 V

208 V

208 V

You will find F, D, and D to ground will be approximately 120 V

F

D

E

G

Figure 12-5: Approximate voltage you will measure on a wye-wye transformer secondary

Practical Wye-Wye Transformer Voltage Measurements

The power distribution system most often used in industrial buildings such as broadcast station studios, office buildings, and the like is a wye-wye connected transformer. The schematic diagram and the voltages you should expect to measure are shown in Figure 12-5. Should you have to troubleshoot the system, you'll find the single-phase 120 volt loads connected between F, D, E, and ground. If there is a 3-phase load, it will be connected to terminals F, D, and E.

Frequently Needed
Troubleshooting Aids

This chapter contains valuable practical troubleshooting aids needed by every electronics technician. Included are charts and graphs to simplify scores of repair problems. All have detailed instructions on their use, and they are designed to save you hours of mathematical calculations. You'll also find work-saving conversion tables, color codes that will help you quickly identify leads when working with transformers, formulas that you may have forgotten, and much more practical, essential, information that you'll need during troubleshooting.

How to Solve Parallel Resistor, Inductor,
and Series Capacitor Problems Without Math

We've all been taught that the combined value of resistors in parallel or capacitors in series is equal to the reciprocal of the sum of the reciprocals of the individual values. You'll probably remember fighting these problems in your first few weeks of electronics school. Even though electronic calculators have made the use of formulas like these fairly simple, it still takes time. And there is a better way. Chart 13-1 will solve even the toughest problems of parallel resistors and series capacitors in a few seconds, and all you need is a simple straightedge to lay across the scales.

When you use the chart you can read scale numbers in ohms,

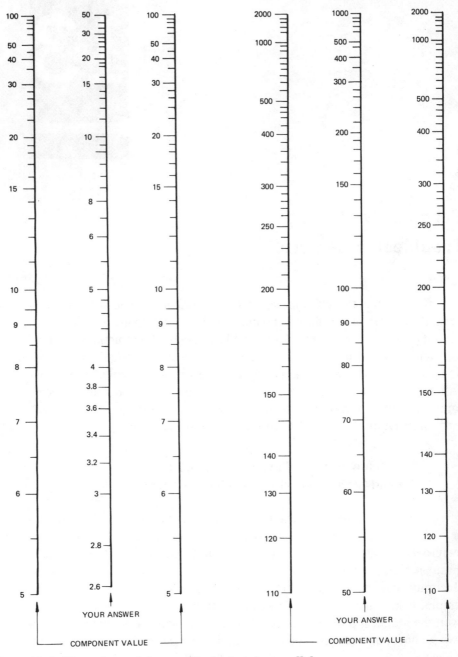

Chart 13-1: Chart for resistors and inductors in parallel and capacitances in series problems

Original idea, H.P. Manly, Radio Electroni

megohms, microfarads, or picofarads, depending on which you're working with, and provided that you read the same units on all scales while solving the problem.

To extend the range of the scales, simply multiply or divide all scales by the same number. For example, if you divide all scales by 10, the two outside scales become 0.5 to 20 and the center scale 0.26 to 10. To see how it works, place a straightedge at 6 and 12 on the outer scales of the left-hand section of the chart. You should read 4 on the center scale. Now, if you are working with two picofarad capacitors in series, your combined answer is 4 picofarads. Or, if two resistors in parallel, the combined answer is 4 ohms.

As another example, if you want a combined value of 4 picofarads and you have a 6 picofarad capacitor on hand, it will take a 12 picofarad capacitor to do the job, as the chart clearly shows. And, to work with three or more resistors or capacitors, it's only necessary to first work the problem for two, as just explained, and then use your answer as a single value with the next component. With this system, there's no limit to the number of components you can parallel or series, depending on what you are working with (resistors or capacitors).

Finally, because parallel inductances are calculated in the same manner as parallel resistances, you can find the resultant value of two or more inductances. However, this is only true providing you don't have any magnetic coupling between the coils. Incidentally, you can also use the chart to determine parallel inductive reactance or capacitive reactance because they follow the same rules as parallel resistance. In other words, since they are both measured in ohms, just lay your straightedge across the chart just as you would with a parallel resistor problem.

Practical Guide to Transformer
Wire Color Coding

Transformer leads are frequently color coded with the EIA standard color code, which really helps the technician when he is troubleshooting. Figure 13-1 shows the colors of the leads of power transformers, IF transformers, and audio output transformers.

If you run into a transformer that is not color coded, one way to check it is to use an ohmmeter to identify the primary windings. These

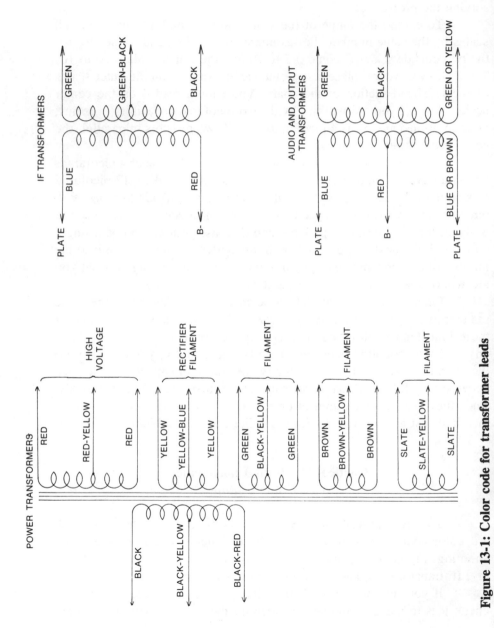

Figure 13-1: Color code for transformer leads

leads are normally black (in some cases, black-yellow) and will have a resistance of anywhere between 5 to 40 ohms. Once you find the primary leads, everything else is easy. Just apply a voltage to the primary and measure the outputs. It's best to apply a low voltage to start off, if possible. If you don't have a low voltage AC source, one trick that works well is to purchase a second-hand electric train speed control transformer. They have a variable AC output and often can be found in very inexpensive places like thrift shops.

As an example of how to make the check, let's assume that you put 10 VAC on the primary and measure 60 volts on the secondary. You have a 6 to 1 step up transformer. This means that placing a 115 VAC on the primary will produce 690 VAC on the secondary in normal operation. A word of caution: If there is any doubt about the transformer having an internal short, or what line frequency it was designed for, always monitor the current in the primary leads during your measurements. If there is a high primary current with no load, you either have an internal short or the transformer was designed to operate on a higher line frequency. For example, many transformers in aircraft equipment operate at 400 Hz and normally cannot be used at 115 V, 60 Hz without severe damage because of the low value of inductive reactance at 60 Hz.

Speaker Output Transformer Color Coding

There are several methods of connecting speaker output transformers. Because of this, the color coding shown in Figure 13-2 can be a great help in identifying speaker output transformer leads while troubleshooting any type of equipment using the systems shown.

Decibel Conversions Using a Straightedge

Even though the conversion of power ratios and voltage and current ratios to decibels can be done by hand, few of us like to cover a sheet of paper with a lot of mathematics. For example, ask someone to tell you quickly how many volts are equal to a −20 dBm. While he's covering that sheet of paper with a lot of mathematics, you can answer the question in a few seconds by using Chart 13-2.

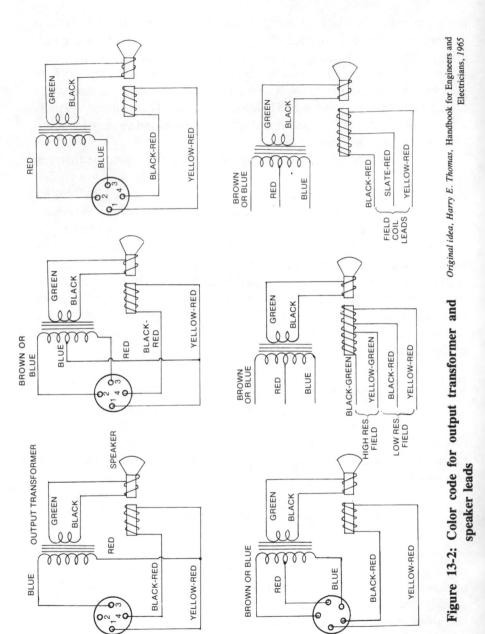

Figure 13-2: Color code for output transformer and speaker leads

Original idea, Harry E. Thomas, Handbook for Engineers and Electricians, 1965

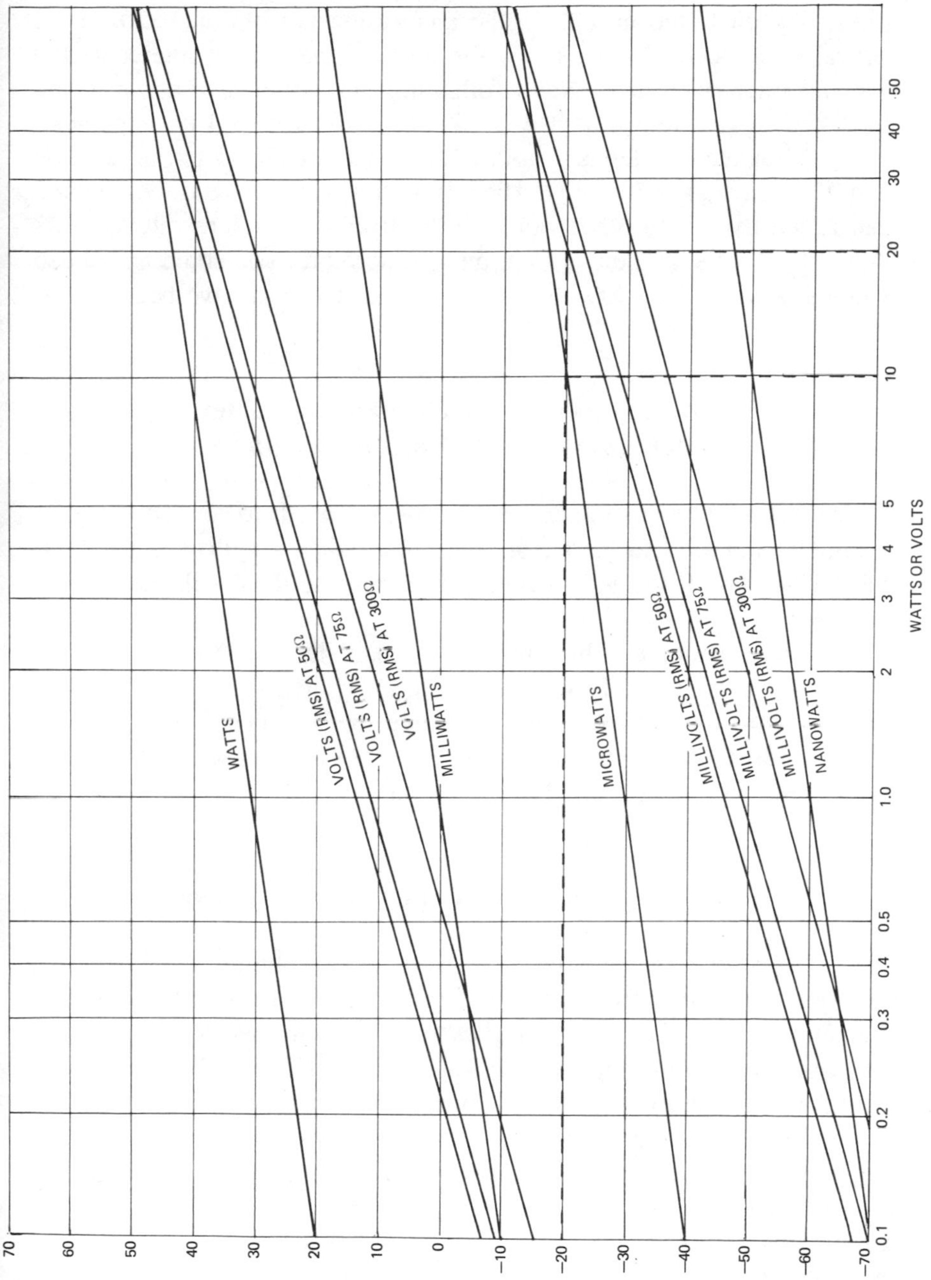

Original idea, Microwaves

Chart 13-2: dBm conversion chart for watts and volts

The first step in making the conversion is to locate −20 on the vertical dBm scale. Next, follow the dashed line until it intercepts the diagonal microwatts line. Now, following the dashed line down to the watts volts scale, we see that it is 10. So, −20 dBm = 10 microwatts.

Your second step is to follow the dashed line over to the junction at millivolts (rms) at 50 ohms. Then drop down with the dashed line and read 22 millivolts. In other words, −20 dBm = 10 microwatts or −22 millivolts. That's all there is to it, and furthermore, you could have been working with a 75 or 300 ohm problem and it would have been just as easy.

An Easy-to-Use and Time-Saving Table
for Conversion of Units of Length

At one time or another every technician needs to convert inches to centimeters, meters to yards, or some other such length measurement. Table 13-1 has all the information you'll need to deal with these problems.

As an example of how to use this chart, let's say that you wish to convert 42.25 meters to feet. Referring to the chart, you will see that it is necessary to multiply by 3.281 to change to feet. Therefore, 3.281 × 42.25 = 138.62 feet. For an even easier way to convert selected frequencies to feet and meters, see Table 13-4 on page 197.

A Simplified Table of Area Conversion Factors

Sometimes when you are constructing a chassis or cabinet for electronics gear, it is important to be able to convert from one unit of area to another, such as from square inches to square centimeters. Or, you may need to convert from square feet to square meters. Table 13-2 removes all doubts about how to make these kinds of conversions.

In dealing with problems of space energy flow, the answer is frequently given in energy per square meter. Therefore, it is often desirable to convert from square meters to square feet. For example, let's assume that you wish to convert 6 square meters to square feet. The table indicates that you must multiply square meters by 10.76 to convert to square feet. Doing this, we get 10.76 × 6 = 64.56 square feet. As you

MULTIPLY NUMBER OF → / BY ↗ / TO OBTAIN ↓	CENTIMETERS	FEET	INCHES	KILOMETERS	NAUTICAL MILES	METERS	MILS	MILES	MILLIMETERS	YARDS
CENTIMETERS	1	30.48	2.540	10^5	1.853×10^5	100	2.540×10^{-3}	1.609×10^5	0.1	91.44
FEET	3.281×10^{-2}	1	8.333×10^{-2}	3281	6080.27	3.281	8.333×10^{-5}	5280	3.281×10^{-3}	3
INCHES	0.3937	12	1	3.937×10^4	7.296×10^4	39.37	0.001	6.336×10^4	3.937×10^{-2}	36
KILOMETERS	10^{-5}	3.048×10^{-4}	2.540×10^{-5}	1	1.853	0.001	2.540×10^{-8}	1.609	10^{-6}	9.144×10^{-4}
NAUTICAL MILES		1.645×10^{-4}		0.5396	1	5.396×10^{-4}		0.8684		4.934×10^{-4}
METERS	0.01	0.3048	2.540×10^{-2}	1000	1853	1		1609	0.001	0.9144
MILS	393.7	1.2×10^4	1000	3.937×10^7		3.937×10^4	1		39.37	3.6×10^4
MILES	6.214×10^{-6}	1.894×10^{-4}	1.578×10^{-5}	0.6214	1.1516	6.214×10^{-4}	2.540×10^{-2}	1	6.214×10^{-7}	5.682×10^{-4}
MILLIMETERS	10	304.8	25.40	10^6		1000			1	914.4
YARDS	1.094×10^{-2}	0.3333	2.778×10^{-2}	1094	2027	1.094	2.778×10^{-5}	1760	1.094×10^{-3}	1

Table 13-1: A simplified table of conversion factors for frequently needed units of length

MULTIPLY NUMBER OF → BY → (columns); TO OBTAIN ↑ (rows)

TO OBTAIN \ MULTIPLY NUMBER OF	ACRES	CIRCULAR MILS	SQUARE CENTIMETERS	SQUARE FEET	SQUARE INCHES	SQUARE KILOMETERS	SQUARE METERS	SQUARE MILES	SQUARE MILLIMETERS	SQUARE YARDS
ACRES				2.296×10^{-5}		247.1	2.471×10^{-4}	640		2.066×10^{-4}
CIRCULAR MILS		1	1.973×10^{5}	1.833×10^{8}	1.273×10^{6}		1.973×10^{9}		1973	
SQUARE CENTIMETERS		5.067×10^{-6}	1	929.0	6.452	10^{10}	10^{4}	2.590×10^{10}	0.01	8361
SQUARE FEET	4.356×10^{4}		1.076×10^{-3}	1	6.944×10^{-3}	1.076×10^{7}	10.76	2.788×10^{7}	1.076×10^{-5}	9
SQUARE INCHES	$6{,}272{,}640$	7.854×10^{-7}	0.1550	144	1	1.550×10^{9}	1550	4.015×10^{9}	1.550×10^{-3}	1296
SQUARE KILOMETERS	4.047×10^{-3}		10^{-10}	9.290×10^{-8}	6.452×10^{-10}	1	10^{-6}	2.590	10^{-12}	8.361×10^{-7}
SQUARE METERS	4047		0.0001	9.290×10^{-2}	6.452×10^{-4}	10^{6}	1	2.590×10^{6}	10^{-6}	0.8361
SQUARE MILES	1.562×10^{-3}		3.861×10^{-11}	3.587×10^{-8}		0.3861	3.861×10^{-7}	1		3.228×10^{-7}
SQUARE MILLIMETERS		5.067×10^{-4}	100	9.290×10^{4}	645.2	10^{12}	10^{6}		1	8.361×10^{5}
SQUARE YARDS	4840		1.196×10^{-4}	0.1111	7.716×10^{-4}	1.196×10^{6}	1.196	3.098×10^{6}	1.196×10^{-6}	1

Table 13-2: A simplified table of area conversion factors

can see, this table makes short work of problems like this one, especially if you have even the simplest of hand-held calculators.

A Helpful Table of Volume Conversion Factors

Table 13-3 is particularly handy when you are designing any type of container where it is necessary to convert from one cubic unit to another. For example, if you want to construct a tank for a certain amount of liquid, you can quickly determine exactly how much cubic space you'll need by using Table 13-3.

If your need is for a cleaning solvent tank that you would like to hold twenty gallons of solvent, you'll first have to convert from gallons liquid to cubic feet. As the chart shows, you must multiply the amount of gallons by 0.1337 to make the conversion. Now, $20 \times 0.1337 = 2.674$ cubic feet. Next, to give yourself a splash board around the top of the tank, increase this value to 3 cubic feet. After doing this, you can see that a tank 1 foot deep and 3 feet long will do the job nicely with about a 4-inch splash guard.

Frequency to Meters or Feet Conversion Table

Frequently, the technician, experimenter, or serviceman must convert from frequency to wavelength in meters or feet or vice versa. Table 13-4 can be used to convert to the desired value anywhere in the frequency spectrum merely by moving the decimal to the right or left.

For example, merely move the decimal to the right one place for all given values of frequency and to the left one place for all listed values of meters and feet, and the table would read from 10 MHz to 500 MHz. Or, move the decimal two places and the table reads from 100 MHz to 5000 MHz. To convert frequencies not listed, simply use Table 13-1.

A Conversion Table for the Most Frequently Needed Basic Units

Table 13-5 can save a lot of time when you run into a conversion problem concerning any of the many units listed. To use the table, simply

TO OBTAIN ↓ / MULTIPLY NUMBER OF → BY	CUBIC CENTIMETERS	CUBIC FEET	CUBIC INCHES	CUBIC METERS	CUBIC YARDS	GALLONS (LIQUID)	LITERS	PINTS (LIQUID)	QUARTS (LIQUID)
CUBIC CENTIMETERS	1	2.832×10^4	16.39	10^6	7.646×10^5	3785	1000	473.2	946.4
CUBIC FEET	3.531×10^{-5}	1	5.787×10^{-4}	35.31	27	0.1337	3.531×10^{-2}	1.671×10^{-2}	3.342×10^{-2}
CUBIC INCHES	6.102×10^{-2}	1728	1	6.102×10^4	46,656	231	61.02	28.87	57.75
CUBIC METERS	10^{-6}	2.832×10^{-2}	1.639×10^{-5}	1	0.7646	3.785×10^{-3}	0.001	4.732×10^{-4}	9.464×10^{-4}
CUBIC YARDS	1.308×10^{-6}	3.704×10^{-2}	2.143×10^{-5}	1.303	1	4.951×10^{-3}	1.308×10^{-3}	6.189×10^{-4}	1.238×10^{-3}
GALLONS (LIQUID)	2.642×10^{-4}	7.481	4.329×10^{-3}	264.2	202.0	1	0.2642	0.125	0.25
LITERS	0.001	28.32	1.639×10^{-2}	1000	764.6	3.785	1	0.4732	0.9464
PINTS (LIQUID)	2.113×10^{-3}	59.84	3.463×10^{-2}	2113	1616	8	2.113	1	2
QUARTS (LIQUID)	1.057×10^{-3}	29.92	1.732×10^{-2}	1057	807.9	4	1.057	0.5	1

Table 13-3: Volume conversion factors for shop use

FREQ. MHz	WAVELENGTH		FREQ. MHz	WAVELENGTH	
	METERS	FEET		METERS	FEET
1.0	300.0	984.0	12.0	25.0	82.0
1.5	200.0	656.0	112.5	24.0	78.72
2.0	150.0	492.0	13.0	23.07	75.69
2.5	120.0	393.6	13.5	22.22	72.88
3.0	100.0	328.0	14.0	21.42	70.28
3.5	85.71	282.14	14.5	20.68	67.86
4.0	75.0	246.0	15.0	20.0	65.6
4.5	66.66	218.66	15.5	19.35	63.48
5.0	60.0	196.8	16.0	18.75	61.5
5.5	54.54	178.9	16.5	18.18	59.63
6.0	50.0	164.0	17.0	17.64	57.88
6.5	46.15	151.38	17.5	17.14	56.22
7.0	42.85	140.57	18.0	16.66	54.66
7.5	40.0	131.2	18.5	16.21	53.18
8.0	37.5	123.0	19.0	15.78	51.78
8.5	35.29	115.76	19.5	15.38	50.46
9.0	33.33	109.33	20.0	15.0	49.20
9.5	31.57	103.57	25.0	12.0	39.36
10.0	30.0	98.4	30.0	10.0	32.80
10.5	28.57	93.71	35.0	8.57	28.11
11.0	27.27	89.45	40.0	7.5	24.6
11.5		85.56	50.0	6.0	19.68

Table 13-4: Frequency to meters or feet conversion table

To Convert	To	Multiply by	Conversely Multiply by
ampere turns	gilberts	1.257	0.7958
British thermal units	kilowatt hours	2.928×10^{-4}	3415
BTU per min	kilowatts	0.01757	56.92
Celsius	Fahrenheit	$(1.8°C) + 32$	$(°F - 32) \times 0.555$
Celsius	Kelvin	$°C + 273.16$	$°K - 273.16$
centimeters	inches	0.3937	2.540
centimeters	meters	0.01	100
centimeters	millimeters	10	0.1
circular mills	square centi-meters	5.067×10^{-6}	1.973×10^5
cubic inches	cubic centi-meters	16.39	0.06102
cubic feet	cubic meters	0.02932	35.31
degrees (angle)	minutes	60	0.01666
degrees (angle)	radians	0.06745	57.3
degrees (angle)	seconds	3600	2.777×10^{-4}
dynes	grams	0.00102	980.392
ergs	joules	10^{-7}	10^7
feet	meters	0.3048	3.2808
gauss	lines per sq. inch	6.452	0.155
gilberts	ampere-turns	0.7958	1.25659
grams	kilograms	0.001	1000
grams	ounces (avoirdupois)	0.03527	28.3527
horse power	watts	745.7	0.001341
joules	ergs	10^7	10^{-7}
kilograms	pounds (avoirdupois)	2.2046	0.453597
kilometers	feet	3281	3.047×10^{-4}
kilometers	meters	10^3	10^{-3}
kilometers	miles (statute)	0.6214	1.60926
liters	gallons (liq. U.S.)	0.2642	3.785
liters	pints (liq. U.S.)	2.113	0.4732
$\log_{10} N$	$\log_{\varepsilon} N$	2.303	
$\log_{\varepsilon} N$	$\log_{10} N$	0.4343	
sq. inches	sq. centimeters	6.452	0.155
sq. feet	sq. meters	0.0929	10.76
webers	gausses (or lines per sq. centimeter)	1.550×10^7	0.645×10^{-7}
webers	maxwells (or lines)	10^8	10^{-8}

Table 13-5: Frequently needed conversion factors

find the known unit in either the first or second column and use the multiplier indicated.

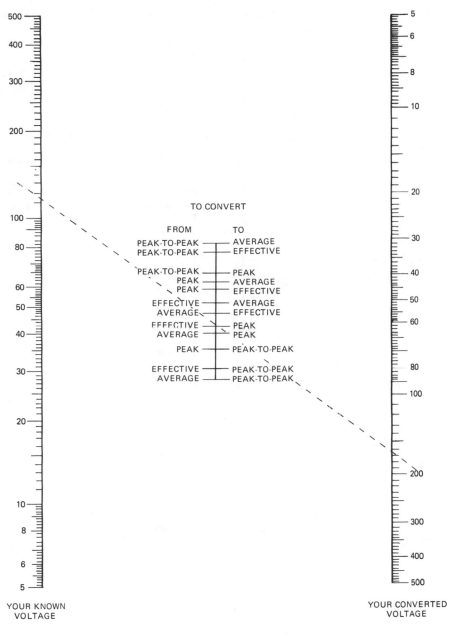

TO CONVERT

FROM	TO
PEAK-TO-PEAK	AVERAGE
PEAK-TO-PEAK	EFFECTIVE
PEAK-TO-PEAK	PEAK
PEAK	AVERAGE
PEAK	EFFECTIVE
EFFECTIVE	AVERAGE
AVERAGE	EFFECTIVE
EFFECTIVE	PEAK
AVERAGE	PEAK
PEAK	PEAK-TO-PEAK
EFFECTIVE	PEAK-TO-PEAK
AVERAGE	PEAK-TO-PEAK

YOUR KNOWN
VOLTAGE

YOUR CONVERTED
VOLTAGE

Chart 13-3: Chart for converting sine wave voltages and currents

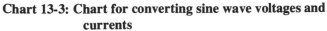

Original idea, Donald Moffat, Electronics World

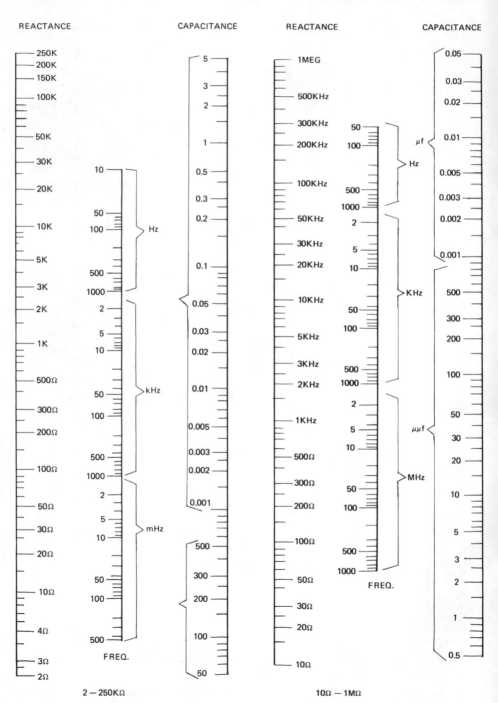

Chart 13-4: Capacitance-to-reactance or reactance-to-capacitance conversion chart

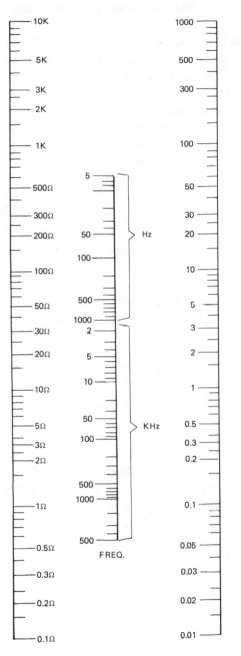

REACTANCE

CAPACITANCE μf

0.2 – 10KΩ

Original idea, H.P. Manly, Radio Electronics

How to Make Sine Wave Voltage and Current
Conversion with a Straightedge

You will find that converting sine wave voltages and currents to average rms (effective), peak, and peak-to-peak values can be almost instantaneous by just using a straightedge and Chart 13-3.

As an example of how to use the chart, suppose that you wanted to know the peak value of the 115 volt (rms) commercial line voltage at your workbench. From 115 on the left hand scale, draw a straight line through rms-peak on the center scale and extend the line to cross the right hand scale, as shown by the dashed line on the chart. At the crossing, read the equivalent peak value of approximately 163 volts. The calculated value is 162.2 volts, so you can see we're only off 0.4 of a volt, which is close enough for most shop work. And another beautiful thing about the chart is that it can be extended indefinitely simply by moving the decimal point the same direction and same number of places on both scales.

Quick and Easy Capacitance-to-Reactance,
Reactance-to-Capacitance Conversion

Following signals through electronic circuits during troubleshooting isn't very hard until you come to an RC network. In other words, when a signal current or voltage divides between a resistor and a capacitor, which way will most of it go? It isn't always that easy to tell because it depends on the value of resistance in one path and capacitive reactance in the other.

Now, it would be nice if all capacitors were marked with a reactance value. However, this can't be done because, as we all know, reactance varies with frequency. There are two ways to beat this problem. 1) Calculate the reactance value by putting the capacitance and frequency into the reactance formula or, 2) a much easier way, use Chart 13-4.

All that is required to determine reactance values is that you lay a straightedge at capacitance on the right-hand scale, at the operating frequency on the center scale, and then read the reactance where the straightedge crosses the left-hand scale. Or if you want to know what value of capacitance to use to provide a certain amount of reactance at a

given frequency, merely place your straightedge on the scale for reactance and frequency, then read the capacitance value on the capacitance
scale. To determine the frequency that you would need to produce a
desired reactance with a known capacitor, put your straightedge on the
known capacitance and reactance.

A Collection of Formulas

Frequently Needed in the Shop

Ohm's law (DC circuits)

$$I = E / R = \sqrt{P / R}$$
$$E = I R = P / I = \sqrt{P R}$$
$$R = E / I = P / I^2 = E^2 / P$$
$$P = I^2 R = E I = E^2 / R$$

Ohm's law (AC circuits)

$$I = E / Z = \sqrt{P / (Z \cos \theta)} = P / E \cos \theta$$

$$E = I Z = P / I \cos \theta = \sqrt{P Z / \cos \theta}$$

$$Z = E / I = P / (I^2 \cos \theta) = (E^2 \cos \theta) / P$$

$$P = I^2 Z \cos \theta = I E \cos \theta = (E^2 \cos \theta) / Z$$

Conductance, susceptance, and admittance formulas

conductance $G = 1 / R$ (for DC circuit). Your answer will be in mhos.
$\qquad\qquad\;\; G = 1 / (R^2 + X^2)$ (for AC circuit). Answer is in mhos.
susceptance $B = 1 / X_C$ (when resistance $= 0$). Answer is in mhos.
$\qquad\qquad\;\; B = 1 / X_L$ (when resistance $= 0$). Answer is in mhos.
admittance $Y = 1 / \sqrt{R^2 + X^2}$ Answer is in mhos.

Capacitance and inductances when reactance is known

$C = 1 / 2 \pi F X_C \qquad$ and
$L = X_L / 2 \pi F$

Power factor formulas (pf)

$pf = \cos \theta$, where θ is the angle of lead or lag

$pf = $ watts $/$ (current) (voltage)

Inductance and capacitance needed for resonance when frequency and one value are known.

$L = 1 / (4 \pi^2) (F_R^2) C \qquad$ and
$C = 1 / (4 \pi^2) (F_R^2) L$

Impedance formulas

$Z \, \underline{/0} = \sqrt{R^2 + X^2} \; \tan^{-1} X/R$ (for series curcuits)
$Z/\theta = R X / \sqrt{R^2 + X^2} \; \tan^{-1} R / X$ (for R and X in parallel)

Decibel formulas

$dB = 10 \log P_2 / P_1 = 20 \log E_2 / E_1 = $
$20 \log I_2 / I_1$ (when input-output impedances are equal)

$dB = 10 \log P_2 / P_1 = $
$20 \log E_2 \sqrt{Z_1} / E_1 \sqrt{Z_2} = $
$20 \log I_2 \sqrt{Z_2} / I_1 \sqrt{Z_1}$

Transformer voltage, current, and turns ratio

$$\frac{\text{voltage on the primary}}{\text{voltage on the secondary}} = \frac{\text{current in the secondary}}{\text{current in the primary}} = \frac{\text{turns of the primary}}{\text{turns of the secondary}}$$

considering the transformer to be 100% efficient and having unity coupling

Transformer impedance ratio

$$\left(\frac{\text{number of turns on the primary } ^{-2}}{\text{number of turns on the secondary} ^{-2}}\right) = \frac{\text{impedance of the primary}}{\text{impedance of the secondary}}$$

Power supply voltage regulation

$$\text{regulation} = \frac{\text{no load voltage } - \text{ full load voltage}}{\text{full load voltage}}$$

no load voltage = (full load voltage)(regulation + 1)

full load voltage = (no load voltage) / (regulation + 1)

Calculating the size of a conductor required for a known load and distance

circular mills = 21.6 × current × distance in feet divided by 2

Frequency modulation index

modulation index = frequency deviation / modulation frequency

Antenna length

$$1/2 \text{ wavelength (narrow band)} = \frac{492 \times 0.95 \text{ (feet)}}{\text{freq in MHz}}$$

$$1/2 \text{ wavelength (wide band)} = \frac{468}{\sqrt{(\text{freq 1}) (\text{freq 2})}}$$

Field strength vs distance

The strength of a radiated field varies inversely with the square of the distance.

$$\frac{\text{old field strength}}{\text{new field strength}} = \frac{\text{new distance}}{\text{old distance}}$$

$$\left(\frac{\text{new distance } ^{-2}}{\text{old distance}}\right) = \frac{\text{new power}}{\text{old power}}$$

Index